ENVIRONMENTAL HEALTH - PHYSICAL, CHEMICAL AND BIOLOGICAL FACTORS

URANIUM CONTAMINATION IN THE NAVAJO NATION

BACKGROUND AND CLEANUP EFFORTS

ENVIRONMENTAL HEALTH - PHYSICAL, CHEMICAL AND BIOLOGICAL FACTORS

Additional books in this series can be found on Nova's website under the Series tab.

Additional e-books in this series can be found on Nova's website under the e-book tab.

ENVIRONMENTAL HEALTH - PHYSICAL, CHEMICAL AND BIOLOGICAL FACTORS

URANIUM CONTAMINATION IN THE NAVAJO NATION

BACKGROUND AND CLEANUP EFFORTS

EUGENE STEPP
EDITOR

nova publishers
New York

For permission to use material from this book please contact us:
Telephone 631-231-7269; Fax 631-231-8175
Web Site: http://www.novapublishers.com

Additional color graphics may be available in the e-book version of this book.

Library of Congress Cataloging-in-Publication Data

ISBN: 978-1-63321-678-5

Published by Nova Science Publishers, Inc. † New York

CONTENTS

Preface		**vii**
Chapter 1	Uranium Contamination: Overall Scope, Time Frame, and Cost Information Is Needed for Contamination Cleanup on the Navajo Reservation *United States Government Accountability Office*	**1**
Chapter 2	Health and Environmental Impacts of Uranium Contamination in the Navajo Nation: Five-Year Plan *Bureau of Indian Affairs, Department of Energy, Nuclear Regulatory Commission, Environmental Protection Agency and Indian Health Service*	**69**
Chapter 3	Federal Actions to Address Impacts of Uranium Contamination in the Navajo Nation: Five-Year Plan Summary Report *Bureau of Indian Affairs, Department of Energy, Nuclear Regulatory Commission, Environmental Protection Agency, and Indian Health Service*	**101**
Index		**149**

PREFACE

This book examines the extent to which the agencies achieved the targets set in the 5-year plan and the reasons why or why not; what is known about the future scope of work, time frames, and costs; and any key challenges faced by the agencies in completing this work and any opportunities to overcome them.

Chapter 1 – Four million tons of uranium ore were extracted from mines on the Navajo reservation primarily for developing the U.S. nuclear weapons stockpile. For over 30 years, the Navajo people have lived with the environmental and health effects of uranium contamination from this mining. In 2008, five federal agencies adopted a 5-year plan that identified targets for addressing contaminated abandoned mines, structures, water sources, former processing sites, and other sites. Federal agencies also provide funding to Navajo Nation agencies to assist with the cleanup work.

GAO was asked to examine the agencies' cleanup efforts. This report examines (1) the extent to which the agencies achieved the targets set in the 5-year plan and reasons why or why not; (2) what is known about the future scope of work, time frames, and costs; and (3) any key challenges faced by the agencies in completing this work and any opportunities to overcome them. GAO examined agency documents; interviewed agency officials, tribal leaders, and stakeholders; and visited sites on the Navajo and Hopi reservations.

Chapter 2 – Mine operators extracted nearly four million tons of uranium ore from 1944 to 1986 under lease agreements with the Navajo Nation. As a result, uranium mining has left the Navajo Nation with a legacy of over 500 abandoned uranium mines (AUMs), four inactive uranium milling sites, a former dump site, contaminated groundwater, structures that may contain elevated levels of radiation, and environmental and public health concerns.

Chapter 3 – The Navajo Nation encompasses more than 27,000 square miles, spread between three states in the Four Corners area. The unique geology of the region makes the Navajo Nation rich in uranium, a radioactive ore in high demand after the development of atomic power and weapons at the close of World War II in the 1940s. Approximately four million tons of uranium ore were extracted during mining operations within the Navajo Nation from 1944 to 1986. Many Navajo people worked the mines, often living and raising families in close proximity to the mines and mills.

Uranium mining and milling activities no longer occur on Navajo lands, but the legacy of these activities remains; including more than 500 abandoned uranium mine claims with thousands of mine features such as pits, trenches, holes, etc., and homes and drinking water sources with elevated levels of uranium, radium and other radionuclides. Uranium and other elements (selenium, arsenic, etc.) associated with mine and mill sites also occur naturally at elevated levels in rock, soil, surface water, and groundwater across the Navajo Nation and the broader Four Corners region. Health effects as a result of exposure to these elements can include lung cancer, bone cancer, and impaired kidney function.

In: Uranium Contamination in the Navajo Nation
Editor: Eugene Stepp
ISBN: 978-1-63321-678-5

Chapter 1

URANIUM CONTAMINATION: OVERALL SCOPE, TIME FRAME, AND COST INFORMATION IS NEEDED FOR CONTAMINATION CLEANUP ON THE NAVAJO RESERVATION*

United States Government Accountability Office

WHY GAO DID THIS STUDY

Four million tons of uranium ore were extracted from mines on the Navajo reservation primarily for developing the U.S. nuclear weapons stockpile. For over 30 years, the Navajo people have lived with the environmental and health effects of uranium contamination from this mining. In 2008, five federal agencies adopted a 5-year plan that identified targets for addressing contaminated abandoned mines, structures, water sources, former processing sites, and other sites. Federal agencies also provide funding to Navajo Nation agencies to assist with the cleanup work.

GAO was asked to examine the agencies' cleanup efforts. This report examines (1) the extent to which the agencies achieved the targets set in the 5-year plan and reasons why or why not; (2) what is known about the future scope of work, time frames, and costs; and (3) any key challenges faced by the agencies in completing this work and any opportunities to overcome them. GAO examined agency documents; interviewed agency officials, tribal leaders, and stakeholders; and visited sites on the Navajo and Hopi reservations.

WHAT GAO RECOMMENDS

Congress should consider requiring federal agencies to develop an overall estimate of the remaining scope of work, time frames, and costs to fully address uranium contamination.

* This is an edited, reformatted and augmented version of the United States Government Accountability Office publication, GAO-14-323, dated May 2014.

GAO also recommends that BIA address its project management challenges, and agencies incorporate key practices into their coordinated outreach strategy. In commenting on a draft of this report, the agencies generally agreed with GAO's recommendations.

WHAT GAO FOUND

Federal agencies implementing the 2008 5-year plan, including the Environmental Protection Agency (EPA), the Department of Energy, the Bureau of Indian Affairs (BIA), and the Indian Health Service, met the targets in six of the plan's eight objectives, working in cooperation with tribal agencies, including the Navajo Nation Environmental Protection Agency. Reasons agencies met the targets were primarily because additional federal and other resources were dedicated to these efforts compared with prior years. For example, from 2008 through 2012, EPA spent $22 million to test and replace contaminated houses, compared with $1.5 million spent in the preceding 5 years. In contrast, targets for two objectives—cleanup of the Northeast Church Rock mine and Tuba City Dump—were not met primarily because EPA's and BIA's estimated schedules were optimistic, and EPA added additional work that extended the time frames. BIA experienced project and contract management challenges in conducting work at Tuba City Dump and did not always follow best practices when estimating the schedule for assessment work at the site. These challenges, if not addressed, could affect BIA's ability to meet future targets for cleanup at the site and successfully plan for project resources.

Federal agencies have not identified the full scope of remaining work, time frames, or costs to fully address uranium contamination on or near the Navajo reservation, although they recognize that significant work remains. In 2008, congressional decision makers requested the agencies provide an overall estimate of the full scope of work needed to address the contamination. The 5-year plan the agencies developed in response to this request does not provide a comprehensive estimate; instead, it focuses on the highest priorities over 5 years. EPA officials said that they typically do not provide cost or schedule estimates until a specific cleanup action is selected and that a number of current uncertainties make developing such an estimate difficult. Even with significant uncertainties, GAO has reported that agencies can create high-level estimates of costs and time frames that can be useful for decision makers and stakeholders. The agencies have collected important information that could provide a starting point for such an estimate. However, absent a statutory requirement to develop such a comprehensive estimate, it appears unlikely that the agencies will undertake such an effort. As a result, decision makers and stakeholders will not have the information they need to assess the overall pace of the cleanup efforts or make resource allocation decisions.

Federal agencies face a variety of challenges in continuing to address uranium contamination on or near the Navajo reservation. For example, according to EPA officials, funding for EPA's efforts at the Navajo abandoned uranium mines is expected to decrease from funding levels available during the 2008 5-year plan because of overall declining federal resources for cleanup. Further, agencies face challenges in effectively engaging tribal communities, in part, because agencies have not always collaborated on their outreach efforts. These agencies identified opportunities to enhance their collaboration by creating a coordinated outreach strategy for the next 5-year plan. Creating such a strategy is consistent with one of the several key practices that GAO has reported can enhance and sustain interagency collaboration and help ensure that agencies make efficient use of limited resources.

ABBREVIATIONS

ATSDR	Agency for Toxic Substances and Disease Registry
AUM	abandoned uranium mine
BIA	Bureau of Indian Affairs
DOE	Department of Energy
CDC	Centers for Disease Control and Prevention
CERCLA	Comprehensive Environmental Response, Compensation, and Liability Act
EPA	Environmental Protection Agency
IHS	Indian Health Service
NNEPA	Navajo Nation Environmental Protection Agency
NPL	National Priorities List
NRC	Nuclear Regulatory Commission
OMB	Office of Management and Budget
RI/FS	remedial investigation and feasibility study
UMTRA	Navajo Uranium Mill Tailings Remedial Action program
UMTRCA	Uranium Mill Tailings Radiation Control Act of 1978

May 5, 2014

Congressional Requesters

Beginning in the 1940s and continuing for approximately 40 years, private companies extracted approximately 4 million tons of uranium ore from mines on the Navajo reservation. Until 1970, the U.S. government purchased most of the ore to meet the demand for developing the nation's first atomic bomb and the subsequent production of weapons for the U.S. nuclear weapons stockpile. Uranium production directly affected the Navajo people who worked in the mines to extract the uranium ore or who lived and raised families in close proximity to the mines. Active uranium mining on the reservation ceased by 1986, and companies often abandoned the mines. Nearly 30 years later, the Navajo people continue to live with the environmental and health effects from mining operations: more than 500 abandoned mines are located across the reservation, some close to homes and communities, and an unknown number of homes and drinking water sources contain radioactive elements. According to the Environmental Protection Agency (EPA) and other federal agencies, health effects—including lung cancer, bone cancer, and impaired kidney function—can result from exposure to elevated levels of uranium and other radionuclides.[1] Uranium contamination at one location also directly affects the Hopi Tribe.[2]

After uranium mining on the Navajo reservation ended, some federal and tribal agencies began to address different aspects of the legacy of uranium contamination. These efforts on the whole have been sporadic, with some exceptions. The largest effort was conducted by the

Department of Energy (DOE), which completed remedial actions at four uranium mill sites on the Navajo reservation where uranium ore was processed under the Uranium Mill Tailings Radiation Control Act of 1978, as amended (UMTRCA).[3] DOE also began monitoring groundwater contamination or implementing plans to address groundwater contamination at three of these sites. Among other efforts, from the 1970s to the mid-2000s, EPA surveyed 65 structures and replaced 2 that had elevated levels of radiation; federal and tribal agencies and others sampled thousands of unregulated water sources to identify contamination; and a Navajo mine reclamation program addressed many of the physical hazards posed by abandoned mines. Even though these efforts represented some progress, Navajo and Hopi tribal officials have repeatedly expressed the tribes' frustrations with the pace of federal efforts to address uranium-related issues.

In October 2007, the House Committee on Oversight and Government Reform held hearings on the health and environmental effects of uranium contamination on the Navajo reservation. Based, in part, on information received in that hearing, in January 2008, the committee requested that five federal agencies create a 5-year plan to address the effects of uranium contamination on Navajo and Hopi tribal lands. All five agencies were either directly involved with uranium mining and processing or with addressing the environmental and health effects of uranium contamination, or both, and included: EPA, DOE, the Department of the Interior's Bureau of Indian Affairs (BIA), the Department of Health and Human Services' Indian Health Service (IHS), and the Nuclear Regulatory Commission (NRC). This plan—referred to in this report as the 2008 5-year plan—was the first collective plan and included actions organized under eight objectives, each of which was to be carried out by one or more of the federal agencies from 2008 through 2012.[4] The plan focused on addressing, over the 5-year period, what the agencies identified as the most urgent uranium-related problems and was not intended to be a long- term plan for fully resolving the entirety of the contamination. The eight objectives were as follows:

1. **Assess abandoned uranium mines for contamination and begin to identify potentially responsible parties**. Assessments are needed to understand the scope of contamination and to identify necessary cleanup actions, and range from radiation scans to field surveys to more thorough site evaluations. Potentially responsible parties, such as former owners or operators of mines, can be compelled to clean up the contamination or to pay cleanup costs under the Comprehensive Environmental Response, Compensation, and Liability Act (CERCLA).[5]
2. **Clean up the Northeast Church Rock mine**. The Navajo Nation identified this mine, which produced one of the largest volumes of uranium of all the Navajo uranium mines, as its highest priority for cleanup. Located near Church Rock, New Mexico, a former operator of the mine is involved in the cleanup efforts.
3. **Assess and clean up contaminated houses and other structures**. Some Navajo community members used materials from the sites that were contaminated with uranium, such as rocks and pieces of wood or metal, to build homes and other structures.
4. **Test and provide alternatives to contaminated, unregulated drinking water sources**. Federal and tribal agencies have estimated that from 15 to 30 percent of Navajos do not have regulated, piped drinking water in their homes.[6] These residents

typically haul water to their homes, and it is not uncommon for them to use unsafe, unregulated water sources, such as livestock wells, for domestic water use.

5. **Continue to remediate groundwater at former uranium processing sites.** DOE is developing or implementing plans to clean up groundwater contamination found at three of the four former processing sites. DOE also conducts long-term surveillance and maintenance at the sites.
6. **Assess the "Highway 160 site."** This site, located across a highway from one of the former uranium processing sites, consists of an area where waste from the processing site was dumped and buried.
7. **Clean up the Tuba City Dump**. A former dump serviced by BIA, Tuba City Dump is located primarily on Hopi land, but it encompasses an area of Navajo land as well, and was used by the nearby communities for more than 50 years. BIA, EPA, and the tribes have identified high levels of uranium in the groundwater underneath a portion of the landfill, and BIA has entered into an administrative settlement agreement with EPA to assess contamination and potential cleanup at the site.[7]
8. **Assess and treat health conditions**. IHS is responsible for treating health conditions in eligible Indians, including any health conditions that may result from exposure to uranium.

Under each plan objective, the relevant federal agencies established targets that they expected to meet by the end of the 5-year period, in consultation with the Navajo Nation. Although not identified as a specific objective, Members of Congress also requested that the federal agencies work together to ensure a coordinated federal response. During the course of the 5-year plan period, health research efforts by a sixth agency—the Agency for Toxic Substances and Disease Registry (ATSDR)—was rolled into the health objective.[8] The federal agencies published a summary report in January 2013, outlining their accomplishments under the plan.[9]

Since the 2008 5-year plan period ended in 2012, the federal agencies have continued their work and started the process of developing their next plan, which they intend to publish in 2014. During this transition, you asked us to report on the federal agencies' efforts under the 2008 5-year plan and going forward. This report examines (1) the extent to which federal agencies achieved the targets identified in the 2008 5-year plan, and the reasons why or why not; (2) what is known about the scope of work, time frames, and estimated costs of fully addressing uranium contamination; and (3) the key challenges, if any, faced by federal agencies in completing this work and the opportunities, if any, which may be present to help overcome these challenges.

To determine the extent to which federal agencies achieved the targets identified in the 2008 5-year plan, and reasons why they did or did not, we compared the agencies' targets with the actions taken by the agencies and their partners over the 5-year plan period, reviewed key documents, and interviewed federal agency and tribal officials. To identify what is known about the scope of work, time frames, and estimated costs of fully addressing uranium contamination on or near the Navajo reservation, we reviewed available documents and interviewed federal agency and tribal officials. We also assessed the schedule and cost estimates for the Tuba City Dump based on our March 2009 cost estimating and assessment guide, a compilation of cost-estimating best practices drawn from across government and industry, and our May 2012 schedule assessment guide.[10] To ascertain any key challenges faced by federal agencies in completing this work and the opportunities that may be present to help overcome these challenges, we reviewed reports and interviewed federal agency and tribal officials, as well as knowledgeable stakeholders, who included local government officials and community members, university researchers, and representatives of nonprofit organizations. We also visited abandoned uranium mines, a former uranium processing site, and other relevant sites on the Navajo and Hopi reservations; we selected these sites based on the level of activity that federal and tribal agencies conducted there during the 2008 5-year plan or because the agencies identified the sites as needing cleanup in the near future. Appendix I describes our objectives, scope, and methodology in more detail.

We conducted this performance audit from January 2013 to May 2014 in accordance with generally accepted government auditing standards. Those standards require that we plan and perform the audit to obtain sufficient, appropriate evidence to provide a reasonable basis for our findings and conclusions based on our audit objectives. We believe that the evidence obtained provides a reasonable basis for our findings and conclusions based on our audit objectives.

BACKGROUND

In keeping with its trust responsibility with respect to Indian tribes, the federal government holds title to the Navajo and Hopi tribal land in trust for the benefit of the tribes and their members. In this context, this section provides information on (1) the Navajo Nation and Hopi Tribe; (2) uranium mining and processing on the Navajo reservation and its environmental effects; (3) Navajo people's exposure to uranium contamination and related health effects; (4) key statutes relevant to addressing uranium contamination; and (5) the roles of federal and tribal agencies and selected actions taken to address uranium contamination on the Navajo and Hopi reservations prior to 2008.

Navajo Nation and Hopi Tribe

The Navajo reservation consists of more than 24,000 square miles of land—about the size of the state of West Virginia—in the states of Arizona, New Mexico, and Utah, making it the largest reservation, geographically, in the United States. The Hopi reservation consists of approximately 2,500 square miles of land in northeastern Arizona, entirely surrounded by the Navajo reservation. Figure 1 shows the locations of the Navajo and Hopi reservations, as well as the locations of 521 abandoned uranium mines and other key sites.

According to the 2010 Census, 174,000 people lived on Navajo land, and, according to Census Bureau estimates, more than 90 percent of the population identified as Navajo. Navajo culture is historically agrarian, and the Navajo people tend to live in small group clusters that are widely dispersed across the reservation. Many Navajo graze sheep and other livestock, which they use for wool and for consumption, among other things. According to Census Bureau estimates, the Navajo reservation's poverty rate is more than twice as high as the poverty rate in the state of Arizona, with 38 percent of people on the reservation—and 44 percent of all children on the reservation—living in poverty. Most homes do not have electricity or telephones; roads are unpaved, and there are no urban centers, although there are large towns generally found near the boundaries of the reservation. Residents living in homes without piped, regulated water sources haul their water from a nearby source, and many of these sources are unregulated, untreated water sources such as livestock wells or natural springs. The Navajo Nation government includes 110 local government subdivisions, known as chapters.

There are more than 5,000 Hopi living on Hopi land, according to the 2010 Census. Poverty rates on the Hopi reservation are similar to those among the Navajo. There were no uranium mines on the Hopi reservation.

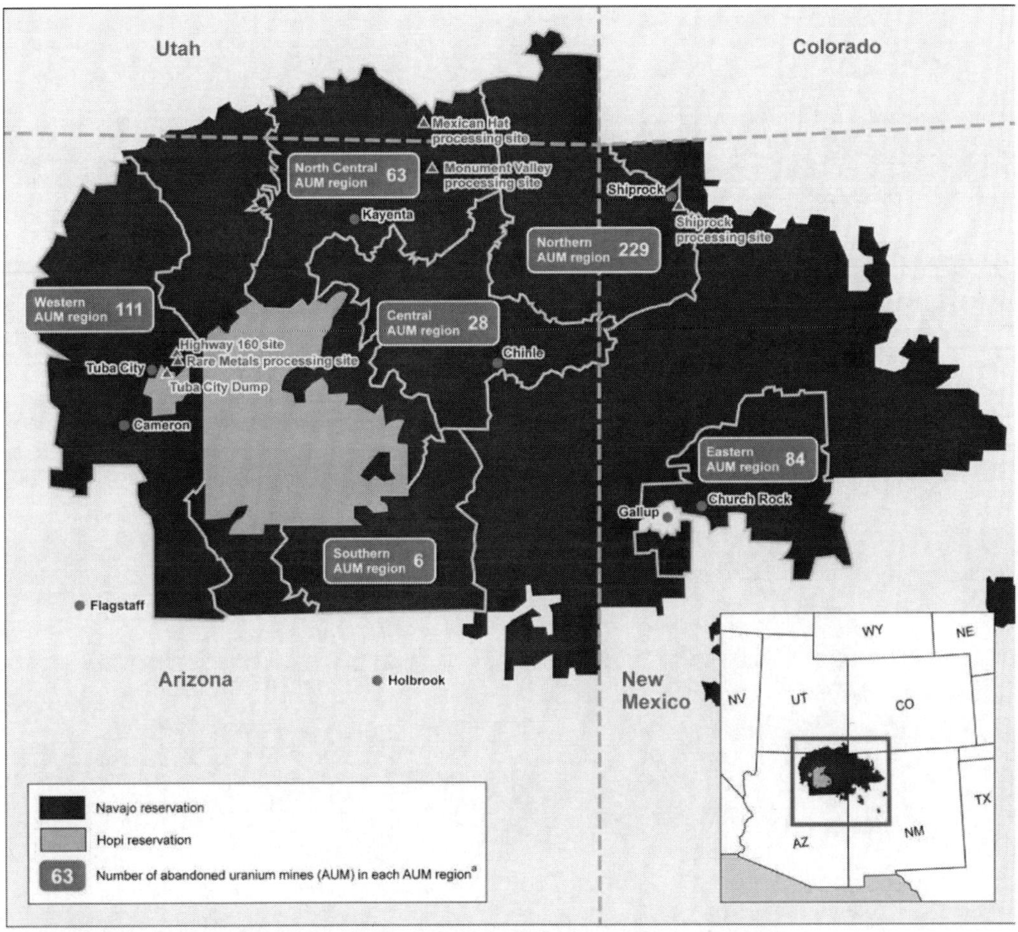

Source: GAO analysis of EPA and DOE data; Map Resource (map).

Note: The eastern portion of the Navajo reservation, as depicted here, is referred to as the Checkerboard area and consists of land with different ownership and statuses, including tribal trust lands; Indian allotments; Navajo tribal fee lands; and private, state, and federal lands.

[a] The Western, North Central, Northern, and Eastern abandoned uranium mine (AUM) regions extend up to 1 mile beyond the borders of the Navajo reservation.

Figure 1. Map of the Navajo and Hopi Reservations with 521 Abandoned Uranium Mines, Four Former Uranium Processing Sites, and Other Key Sites.

Uranium Mining and Processing on the Navajo Reservation and its Environmental Effects

The Navajo reservation is located on the southern end of a stretch of naturally occurring uranium deposits that spans the western United States (see figure 2). The uranium found on the reservation is primarily located in sandstone formations that range from surface outcrops to deposits more than 4,000 feet deep.

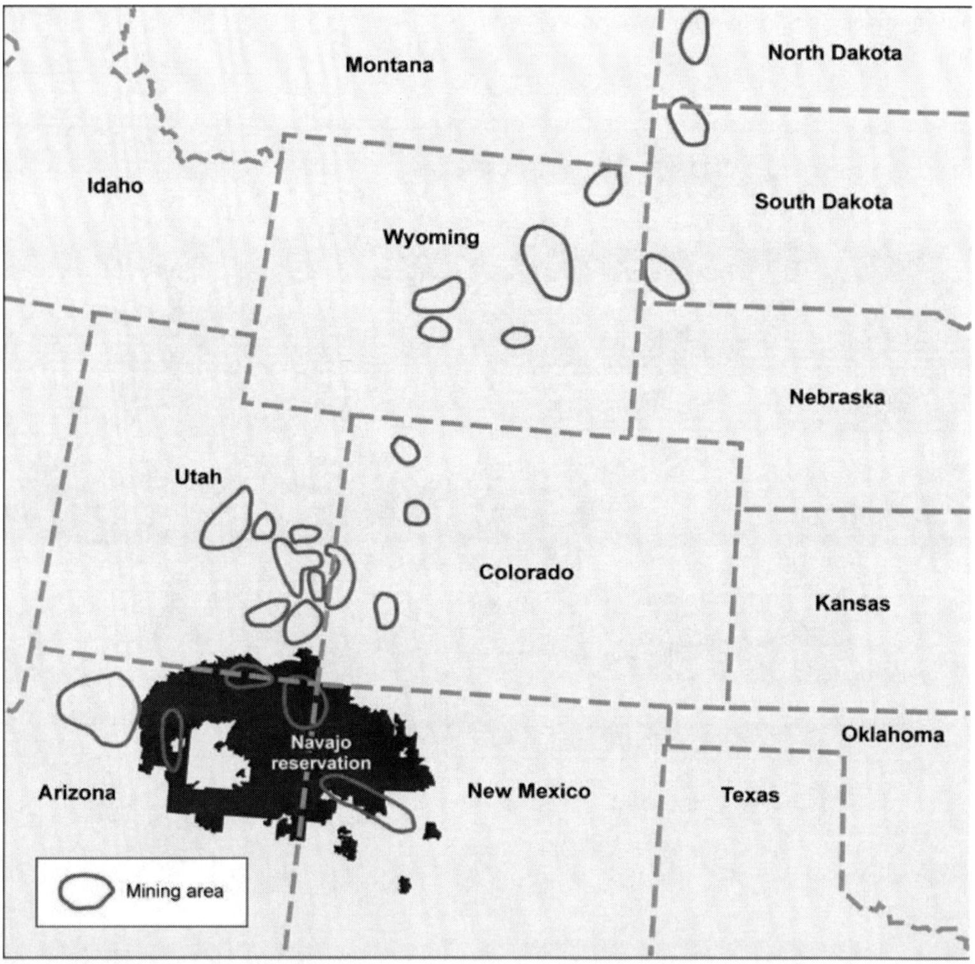

Source: DOE; Map Resources (map).

Notes: The eastern portion of the Navajo reservation, as depicted here, is referred to as the Checkerboard area and consists of land with different ownership and statuses, including tribal trust lands; Indian allotments; Navajo tribal fee lands; and private, state, and federal lands.

According to DOE, Texas, Idaho, Oregon, and Washington also contain significant uranium mining areas.

Figure 2. Uranium Mining Areas in the Rocky Mountain and Colorado Plateau Regions of the United States, Including on the Navajo Reservation.

On the Navajo reservation, uranium ore was removed from the ground at more than 500 mines, generally through open pit mining for ore deposits located relatively close to the surface, or underground mining for deeper deposits. The mines were often located in mountainous areas and consisted of multiple features, such as portals and vertical shafts. The material left behind after the ore was removed—known as waste rock— was then disposed, often in nearby piles, and contained dangerous materials, such as radium, radon, and heavy metals. Once mining ceased at a site, companies often abandoned the mines, leaving the waste rock piles in place without conducting any cleanup or posting signs warning about the dangers of contamination or physical hazards. The extracted ore was sent to an off-site processing facility called a mill. At the mill, the mined uranium ore was crushed, ground, and then fed to a leaching system that produced yellow slurry—called yellowcake—that was further processed for use in nuclear weapons or, as of the mid-1960s, for use in nuclear power plants. The leaching system left a waste product known as mill tailings that retained some toxic contaminants. The tailings were of a sandy consistency and mixtures of tailings and water were placed in unlined evaporation ponds at the mill site. DOE estimates that millions of gallons of water contaminated by mill tailings were released into the groundwater over the life of the sites through the unlined ponds. In addition, on July 16, 1979, the largest release of radioactive materials in the United States occurred when a dam on one of the evaporation ponds broke at a processing site near Church Rock, New Mexico, resulting in the release of 94 million gallons of radioactive waste to the Puerco River, which flowed through nearby communities. Figure 3 depicts the uranium mining and processing that occurred on or near the Navajo reservation.

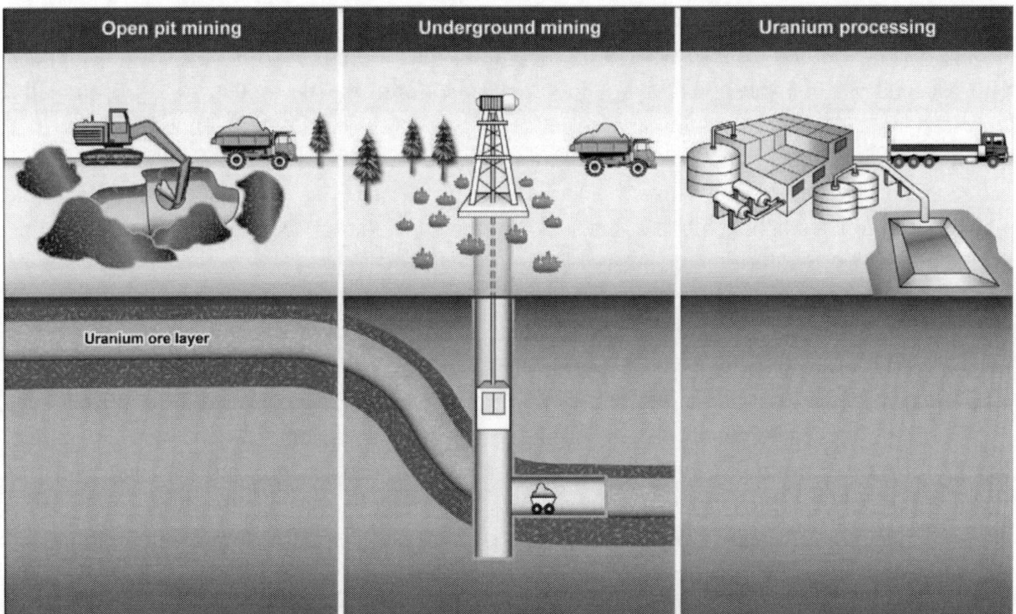

Source: GAO; Art Explosion (tress and heavy equipment images).

Figure 3. Uranium Mining and Processing That Occurred on or Near the Navajo Reservation.

Most of the uranium mining and processing on the Navajo reservation occurred from the late 1940s through the 1960s, and the federal government played a variety of roles during this time. For example, as reported by EPA and the Navajo Nation, beginning in the 1940s, the Secretary of the Interior, along with the Navajo Nation, and later BIA, issued leases and permits to private companies and individuals for uranium mining on the Navajo reservation.[11] In another example, the Atomic Energy Commission (Commission)—a precursor agency to DOE—established a series of financial incentives for the discovery and production of domestic uranium, including guaranteed minimum prices for uranium ore and financial bonuses for uranium ore mined from any previously unidentified site, according to a report prepared by the Commission. According to the report, the Commission also provided infrastructure support, such as roads needed to survey mine sites and transport ore. Finally, the federal government was the sole customer of the processed uranium from 1947 to 1965.

Beginning in 1970, uranium from the Navajo reservation was sold exclusively to the commercial sector for use in nuclear power plants, but prices fell in the 1980s, and uranium mining operations on the reservation ended in 1986. Because of lingering contamination and its effects, in 2005 the Navajo Nation enacted a law placing a moratorium on uranium mining and processing on any site within the tribe's territorial jurisdiction. In 2012, the Navajo Nation enacted a law prohibiting transportation of uranium ore or radioactive waste through lands under the tribe's territorial jurisdiction unless fees, bonding, and other requirements were met. The 2012 law also stated that the Navajo Nation generally opposed transportation of uranium ore or radioactive materials, except for the purpose of disposing of materials from past mining or milling in a long- term facility outside of the tribe's territorial jurisdiction or a temporary facility within the jurisdiction. Even with these laws, however, increases in the price of uranium during the past 10 years have sparked renewed interest in uranium mining and processing on and near the Navajo reservation, and opinions over new mining appear split, especially given the potential for job creation offered by the industry in economically depressed areas.[12] For example, a committee of the Navajo Nation Council approved a resolution in December 2013 acknowledging a private company's right-of-way across tribal land near Church Rock, New Mexico, and authorizing its use for a demonstration project that extracts uranium from beneath the surface. The Navajo Nation Department of Justice has concluded that the resolution conflicts with the 2005 and 2012 laws.

Exposure to Uranium and Its Health Effects

Uranium is present naturally in virtually all soil, rock, and water, and is spread throughout the environment by geological processes, as well as by wind and rain. On the Navajo reservation, winds blow during most of the year, exceeding 50 miles per hour at times, and localized, heavy rainstorms occur throughout the summer. When uranium is present in the environment, people may be exposed to it, as well as the radioactive by-products that are created as uranium decays—including radium and radon (a gas)—through a variety of exposure pathways. For example, Navajo people have been exposed to naturally occurring uranium and its by-products by drinking water from unregulated wells that tap into groundwater that comes into contact with underground uranium deposits. People are also exposed to naturally occurring radon when it migrates into their homes from the uranium-bearing soil underneath.

Mining and milling processes on the Navajo reservation created new pathways for exposure to uranium, increasing the amount of potential exposure at the surface. When the uranium mines were in operation, uranium miners inhaled radioactive dust and radon in and near the mines where they worked. Miners tracked the dust into their homes, exposing their families. Community members occasionally used materials from the mines and mill sites to build their houses and ceremonial structures, leading to increased radon inside these structures. When waste rock piles were left next to abandoned mines, wind and rain at times spread—and could continue to spread—the hazardous materials, sometimes through intermittent streams, where these materials could come into contact with residents of nearby communities, who could then be exposed by inhalation or ingestion. Residents of some communities located near the mines reported playing as children in and around the open mines. Many of the exposure pathways that existed when the mines and processing sites were in operation have been eliminated, and the Navajo Nation has stated that the most significant safety hazards that were present at the abandoned mines have been addressed. However, there are pathways of exposure that remain for Navajo residents today (see figure 4).

Although historic, occupational exposure to uranium has been shown by the Centers for Disease Control and Prevention (CDC) and others to have affected human health, the extent to which the Navajo people have experienced health effects resulting from uranium exposures in other ways has not been thoroughly examined and remains uncertain. For nonoccupational exposures, comprehensive health studies have not been conducted to assess the health effects of uranium contamination on Navajo communities or other communities located near active or abandoned uranium mines and processing sites, but Navajo community members who have lived near these sites have reported a variety of serious health effects, including cancers, according to CDC.[13] EPA reports that exposure to gamma radiation—such as from waste rock located near abandoned mines—can cause a variety of cancers, including lung cancer and leukemia, and that exposure to radon can cause lung cancer. Because of these potential dangers, EPA recommends that people stay away from areas on the Navajo reservation with especially high levels of radiation—more than 10-times above the naturally-occurring, background radiation—in order to avoid potential health effects.[14] ATSDR and EPA have noted that the abandoned mines pose a risk especially to children, since children tend to put dirt in their mouths, and the dirt at the mines could be contaminated. EPA noted in the 2008 5-year plan that inhabitants of structures constructed with uranium mining waste are at risk of developing lung cancer because of the increased presence of radon in indoor air. In addition, given the consumption by Navajo residents of livestock that have grazed on plants located on or near abandoned mine sites, residents and researchers have identified the need to study the potential for exposure to radiation through consuming these animals.

Key Statutes Relevant to Addressing Uranium Contamination

Two key statutes involved with addressing uranium contamination are (1) CERCLA,[15] also known as Superfund, and (2) UMTRCA.[16]

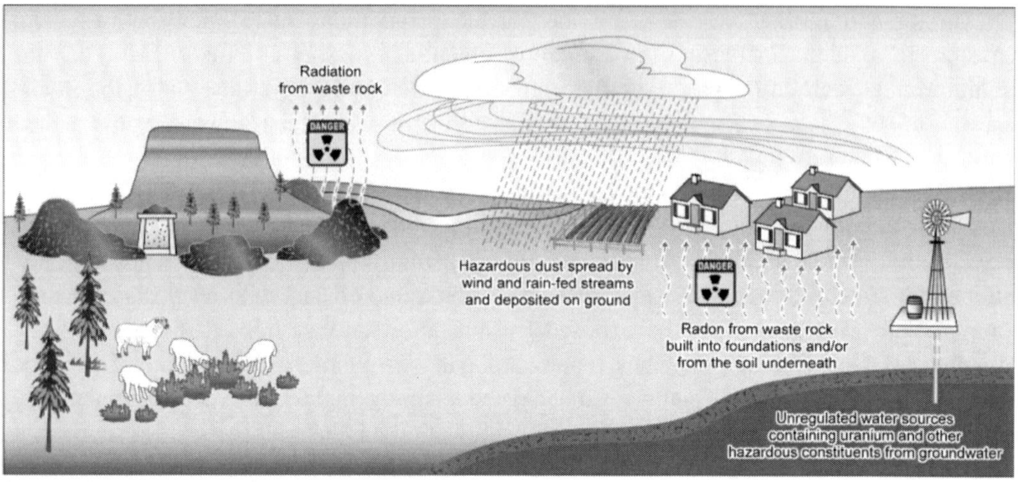

Source: GAO; Art Explosion (tress and sheep).

Figure 4. Current Pathways for Exposure to Uranium and Other Radiation Hazards.

CERCLA

CERCLA established the Superfund program in 1980 to protect human health and the environment from the effects of hazardous substances, including uranium. Under CERCLA, potentially responsible parties—such as current or former owners or operators of a mine site containing hazardous substances—are liable for conducting or paying for cleanup of hazardous substances at contaminated sites.[17] If the federal government is a potentially responsible party at a site, it is liable for cleanup costs even if there are nonfederal potentially responsible parties. For example, one court has held that the federal government was liable as an owner under CERCLA for the cleanup costs at a mine located within an Indian reservation.[18]

Under CERCLA, EPA has the authority to compel potentially responsible parties to clean up contaminated sites, or to conduct cleanups itself and then seek reimbursement from the potentially responsible parties. EPA may compel cleanup by bringing an enforcement action against a potentially responsible party or by attempting to reach an administrative agreement—known as an administrative order on consent, or a settlement agreement—requiring the responsible party to perform and pay for site cleanup. Sometimes, however, potentially responsible parties cannot be identified or may be financially unable to perform the cleanup.

Under the Superfund program, EPA and potentially responsible parties can undertake two types of cleanups: (1) removal actions and (2) remedial actions.[19]

Removal actions are generally shorter-term or emergency cleanups to mitigate immediate threats. These include time-critical removals for threats requiring action within 6 months, and non-time-critical removals for threats where action can be delayed to account for a 6-month or longer planning period, which includes a site evaluation to characterize the site and identify and analyze removal alternatives.[20] Removal actions can include, for example, installing a fence around a contaminated site and excavating contaminated soils for disposal, and can be quite complex. Removal actions can be financed by the Hazardous Substance Superfund Trust Fund (Superfund Trust Fund).[21]

Remedial actions typically are longer-term actions that involve a more elaborate process to permanently and significantly reduce contamination. Remedial actions are taken instead of, or in addition to, a removal action. Before undertaking a remedial action, a remedial investigation and feasibility study (RI/FS) is conducted in accordance with an approved work plan to (1) characterize site conditions and assess the risks to human health and the environment, among other things, and (2) evaluate various options to address the problems identified. A remedy is selected for addressing the site's contamination in a record of decision, and the design of the selected remedy is then developed and implemented (see figure 5). Only sites on the National Priorities List (NPL)—EPA's list of the nation's most contaminated sites—are eligible to have remedial actions financed by the Superfund Trust Fund.[22] None of the mine sites on the Navajo reservation are currently listed on the NPL.

UMTRCA

UMTRCA required DOE to take remedial actions at certain uranium mill sites across the country—and properties in the vicinity that were contaminated with radioactive materials from the mill sites—to stabilize and control the mill tailings in a safe and environmentally sound manner and to minimize or eliminate health hazards, among other things.

UMTRCA included four sites on the Navajo reservation:[23] Mexican Hat, Monument Valley, Shiprock, and Tuba City, which we refer to as the Rare Metals site in this report.[24] Under the act, DOE is also responsible for ensuring that any residual radioactive minerals entering the groundwater do not exceed specified limits; therefore, DOE maintains groundwater remediation systems at sites where groundwater contamination has persisted. In accordance with the act, the Secretary of Energy has entered into cooperative agreements with the Navajo Nation to perform the remedial actions at the sites located on the tribe's land; the current agreement lasts until 2017.[25] The act also required DOE to complete the remedial actions at the processing sites and vicinity properties before DOE's authority under the act, as amended, to perform these actions expired in 1998; DOE's authority to perform groundwater restoration activities has not expired. UMTRCA did not include provisions for DOE to remediate abandoned uranium mines.[26]

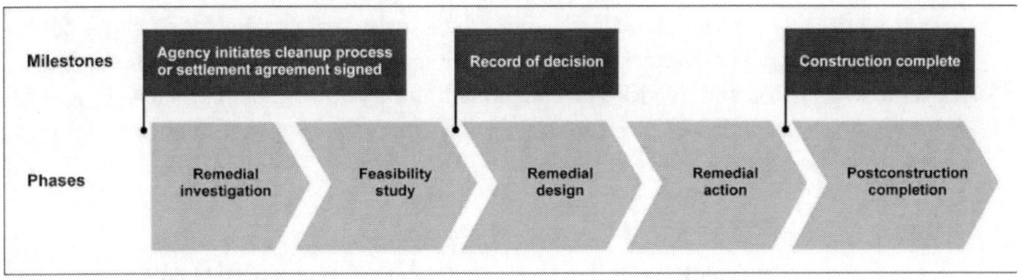

Source: GAO analysis of EPA data.

Figure 5. Superfund Remedial Action Process.

Federal and Tribal Agencies' Roles and Selected Actions Taken to Address Uranium Contamination on or Near the Navajo Reservation Prior to 2008

A variety of federal agencies have specific roles in addressing uranium contamination on or near the Navajo reservation. Table 1 outlines the agencies, their roles, and selected actions they took to begin addressing the contamination prior to 2008.

In addition to the federal agencies, tribal agencies play key roles and have been actively addressing the impacts of historical uranium mining and processing on or near the Navajo reservation.

- *Navajo Nation Environmental Protection Agency (NNEPA).* NNEPA is the lead Navajo agency for regulating radiological contamination at abandoned uranium mines. NNEPA addresses uranium contamination on the Navajo reservation through a variety of programs, including the Navajo Superfund program, which is responsible for assessing hazardous waste sites on the reservation, including abandoned uranium mines. NNEPA partners with EPA in working under the structures, abandoned mines, and unregulated drinking water objectives of the 2008 5-year plan, and provides input on the other objectives as well.
- *Navajo Abandoned Mine Lands Reclamation/ Uranium Mill Tailings Remedial Action (UMTRA) Department.* This Navajo department consists of two programs, the Abandoned Mine Lands Reclamation Program and the UMTRA program. The Abandoned Mine Lands Reclamation Program reclaims abandoned mines on the Navajo reservation.[27] From the 1990s through 2005, the Abandoned Mine Lands Reclamation Program reclaimed more than 900 abandoned uranium mine features found at the 521 mines located on or near the reservation, primarily addressing surface hazards, including stabilizing steep areas and burying uranium-contaminated soils.[28] According to Navajo Nation officials, this work did not address all associated radiological hazards, and the program continues to conduct maintenance on past reclamation work. The department's UMTRA program provides assistance to DOE under an UMTRCA cooperative agreement at the four former uranium processing sites on the reservation.[29] Under the 2008 5-year plan, the Navajo Nation chose not to have the Abandoned Mine Lands Reclamation Program coordinate its reclamation work with EPA and NNEPA's abandoned mine work. The UMTRA program, however, supported DOE's efforts at the former processing sites under the 5-year plan.

Agencies Met the Targets in Six Out of Eight Plan Objectives Primarily Because of Additional Federal Resources

Federal agencies met the targets in six of the eight objectives they established in the 2008 5-year plan, but they did not meet the targets in two of the eight objectives. Reasons the agencies met the targets in five objectives were primarily because additional federal resources, including funding and staff time, were dedicated to their efforts. DOE met the targets in a sixth objective because the agency set targets that represented a continuation of previously required activities. By contrast, the reasons federal agencies did not meet the targets in two objectives were, in part, because of decisions to conduct additional assessment and outreach activities before identifying final cleanup actions. Remaining actions are necessary to meet the targets in these two objectives.

Table 1. Federal Agencies, Their Roles, and Selected Actions Taken to Address Uranium Contamination on or Near the Navajo Reservation Prior to 2008

Agency	Role addressing uranium contamination	Selected actions taken prior to 2008
Environmental Protection Agency (EPA)	EPA Region 9 addresses issues related to contaminated abandoned uranium mines and contaminated homes and other structures. EPA conducts this work under the Comprehensive Environmental Response, Compensation, and Liability Act (CERCLA).[a] EPA also approves BIA's work at the Tuba City Dump under a settlement agreement entered into pursuant to CERCLA and will select the remedy at this site. In addition, EPA funds construction of drinking water infrastructure projects under the Safe Drinking Water Act.[b]	From the 1970s to the 1990s, EPA and the Navajo Nation assessed 65 structures for uranium contamination and replaced 2 contaminated structures. In addition, from 2000 to 2007, EPA worked with Navajo tribal agencies to collect and map information about the abandoned uranium mines on the reservation.[c] Further, in 2007, EPA conducted its first interim removal action at the Northeast Church Rock mine, removing approximately 6,500 cubic yards of contaminated soils from a residential area located adjacent to the mine.
Bureau of Indian Affairs (BIA)	BIA serviced the Tuba City Dump, located near the Navajo town of Tuba City and the Hopi Villages of Moenkopi, which was used by the surrounding communities for 50 years until its administrative closure in 1997. BIA is a potentially responsible party at the site.	In 1998, BIA began assessment activities, including analyses of groundwater and pathways of contaminant migration that were conducted by various consultants and other federal agencies, including the U.S. Geological Survey and EPA. In 2006, BIA initiated the CERCLA remedial action process so that a final remedy could be evaluated, selected, and ultimately implemented at the site.
Department of Energy (DOE)	DOE is responsible, under the Uranium Mill Tailings Radiation Control Act of 1978 (UMTRCA), for long-term surveillance and maintenance at the four former uranium processing sites located on the Navajo reservation.[d] DOE maintains groundwater	DOE completed surface remedial actions at the four former sites from 1986 to 1995, and constructed disposal cells at the Mexican Hat, Shiprock, and the Rare Metals sites (waste from Monument Valley was disposed at Mexican Hat). DOE adopted groundwater compliance action plans for the Shiprock and

Table 1. (Continued)

Agency	Role addressing uranium contamination	Selected actions taken prior to 2008
	remediation systems at three sites. DOE is also the lead federal agency at the Highway 160 site.	Rare Metals sites and began evaluating options at the Monument Valley site to address the continued presence of groundwater contamination. DOE also remediated 31 contaminated properties in the vicinity of former processing sites that were contaminated with radioactive materials from the mill sites.
Nuclear Regulatory Commission (NRC)	NRC regulates and licenses DOE's activities under UMTRCA at the three former uranium processing sites on the Navajo reservation that contain disposal cells and conducts oversight of DOE's monitoring and maintenance activities at all four of the former processing sites.	As required by UMTRCA, NRC evaluated and concurred on DOE's remedial action plans at Mexican Hat, Shiprock, and Rare Metals and then issued licenses to DOE for the custody and long-term care of radioactive disposal cellsat these sites in the 1990s.
Indian Health Service (IHS)	IHS's Navajo Area Indian Health Service provides health care to eligible American Indians through a network of hospitals and health centers. IHS also funds and develops water infrastructure projects to serve Navajo homes.	IHS provided uranium-related health services to a specific set of Navajos, including former uranium miners, former uranium millers, and community members who may have been exposed to fallout from the detonation of nuclear devices.[e]
Centers for Disease Control and Prevention (CDC)/ Agency for Toxic Substances and Disease Registry (ATSDR)	CDC, among other things, conducts scientific research and provides health information to communities. ATSDR was established under CERCLA to carry out that law's health-related activities, which included survey and screening programs to determine the relationships between toxic exposures and illness.[f]ATSDR is located within CDC's Office of Noncommunicable Disease, Injury, and Environmental Health.	CDC was involved in efforts in 2006 and 2007 to sample unregulated drinking water sources and educate Navajo communities about the results. ATSDR did not conduct work related to uranium contamination on the Navajo reservation before 2008.

Sources: GAO analysis of agencies' information.

[a] Pub. L. No. 96-510 (1980), *codified as amended at* 42 U.S.C. §§ 9601-9675.

[b] Pub. L. No. 93-523 (1974), *codified as amended at* 42 U.S.C. §§ 300f to 300j-26.

[c] EPA, et al., *Abandoned Uranium Mines and the Navajo Nation: Navajo Nation AUM Screening Assessment Report and Atlas with Geospatial Data* (August 2007).

[d] Pub. L. No. 95-604, 92 Stat. 3021 (1978).

[e] IHS provided these services to eligible Navajos through the Radiation Exposure Screening and Education Program, which was created in response to the Radiation Exposure Compensation Act Amendments of 2000, which recognized the federal government's responsibility to initiate and support programs designed for the early detection, prevention and education on diseases to aid the thousands of individuals adversely affected by the mining of uranium and testing of the country's nuclear weapons.

[f] Pub. L. No. 96-510, § 104(i), 94 Stat. 2767, 2778-2779 (1980).

Agencies Met the Targets in Six Out of Eight Plan Objectives

In the 2008 5-year plan, federal agencies identified targets under each of the eight objectives that they intended to meet, in cooperation with tribal agency partners, by the end of the plan period in 2012. We found that the agencies met the targets in six of the eight objectives. According to the agencies' January 2013 summary report, the 2008 5-year plan outlined a strategy for gaining a better understanding of the scope of the problem and addressing the greatest risks first. The scope of work required to meet most of the targets did not represent the entirety of work necessary to fully address the issues encompassed by each objective. Table 2 explains the actions taken by the federal agencies—and the tribal agencies with whom they worked—during the period of 2008 through 2012, and our assessment of whether these actions met the targets in the 2008 5-year plan.[30]

We also found that federal agencies completed additional actions and produced results beyond the targets in the 2008 5-year plan during the plan period and in 2013. Among other things, EPA, working with the Navajo Nation, conducted a time-critical removal action at the Skyline mine, located within the Oljato Chapter in southern Utah. The action involved moving 25,000 cubic yards of radioactive mine waste—most of which was located at the bottom of a 700-foot high mesa—to a repository constructed on-site at the top of the mesa. According to EPA officials, the agency built the repository to be permanent, but the waste could ultimately be removed from the site given the Navajo Nation's preference that all contamination be removed from the reservation. EPA undertook smaller, interim removal actions at three other sites, including the Quivira mine, which is located near the Northeast Church Rock mine.[31] At two of the sites, EPA built temporary storage repositories to hold the waste on-site until a final disposal option is selected.

In addition, EPA and NNEPA identified 43 of the 521 abandoned mines as the highest priority for additional assessment work and cleanup actions; EPA officials said these mines are the highest priority because they pose the greatest exposure risks to the Navajo people since elevated radiation levels are present at the mines and they are located near houses or other potentially inhabited structures.[32] EPA recommends that people stay away from areas on the Navajo reservation with such elevated levels of radiation in order to avoid potential health effects. The 43 highest priority mines include 37 mines where radiation levels measured at or above 10-times the background radiation at the mine and where a potentially inhabited house or structure is within one-quarter mile of the mine. Of these mines, 8 mines measured at or above 50-times the background radiation, and 3 mines measured above 95-times the background radiation. EPA and NNEPA identified 6 additional mines as part of the 43 highest priority mines where radiation levels were lower— from 2-times to 10-times background—but that posed especially high risks because, for example, a potentially inhabited house or structure is within 200 feet. EPA Region 9 officials we spoke with said prioritizing the mines benefits both federal and tribal agencies by providing a common road map for their efforts. Figure 6 shows how the 43 highest priority mines relate to the rest of the 521 abandoned uranium mines in terms of radiation levels and distance to homes or structures.

Table 2. Assessment of Actions Taken by Federal Agencies and Their Partners to Address Navajo Uranium Contamination Targets in the 2008 5-year Plan

2008 5-year plan objective	Targets established by federal agencies in 2008	Key actions taken by federal agencies from 2008 through 2012	Lead federal agency (federal and tribal agency partners)	GAO assessment
Abandoned uranium mines—assessment and enforcement	• Conduct initial assessments of up to 250 mines and detailed assessments of approximately 35 mines.[a] • Search for potentially responsible parties.	• The Environmental Protection Agency (EPA) completed initial assessments, which included radiation scans, of 521 mines and more detailed assessments of 45 mines to determine whether the mines could potentially be eligible for listing on the National Priorities List and/or if they should be referred to the agency's emergency response program for immediate action. • EPA continued to search for potentially responsible parties and, by the end of 2012, had signed agreements with potentially responsible parties for 3 mines, and identified potentially responsible parties at 24 additional mines.	EPA (Navajo Nation Environmental Protection Agency (NNEPA))	Met
Northeast Church Rock mine	• Complete cleanup of the former Northeast Church Rock uranium mine.	• EPA and the potentially responsible party conducted two interim removal actions to remove 130,000 cubic yards of contaminated soil found in the surrounding community but did not begin a final cleanup action at the mine. • EPA selected the cleanup remedy for the site and organized an interagency work group to provide technical design assistance, comprised of other federal agencies and others.	EPA (NNEPA, Nuclear Regulatory Commission (NRC), Department of Energy (DOE))	Not met
Contaminated houses and other structures	• In coordination with the Navajo Nation, assess up to 500 structures. • Where contaminated structures are found that pose a health risk, take appropriate response actions using EPA's Superfund authority.	• EPA and NNEPA assessed 878 structures for contamination; EPA then conducted more thorough assessments using more sensitive equipment for 527 structures that posed potential health risks. • EPA identified 43 structures and 18 yards that posed health risks. Of these, EPA demolished and rebuilt or provided financial settlements for 31 structures, excavated 18 yards, and awarded a grant to a Navajo housing agency to rebuild 9 structures.[b]	EPA (NNEPA)	Met

2008 5-year plan objective	Targets established by federal agencies in 2008	Key actions taken by federal agencies from 2008 through 2012	Lead federal agency (federal and tribal agency partners)	GAO assessment
Unregulated drinking water sources	• Test up to 70 unregulated water sources located near abandoned uranium mines to determine if they contain safe levels of radiation or radionuclides and provide results to affected communities. • Provide alternative water supplies where feasible.	• EPA tested 97 unregulated water sources; through this and work completed prior to the 5-year plan, EPA, NNEPA, the Centers for Disease Control and Prevention (CDC), and other partners identified 29 sources with unsafe uranium levels, and worked to communicate the results to communities. • EPA and the Indian Health Service (IHS) funded 13 water infrastructure projects to provide alternative water supplies to communities near more than half of the 29 contaminated sources; EPA funded a pilot project for the Navajo Nation to haul water to remote communities.[c]	EPA (CDC, IHS, NNEPA, Navajo Dept. of Water Resources)	Met
Former uranium processing sites	• Continue existing groundwater treatment strategies at three former uranium processing sites with contaminated groundwater. • Continue long-term surveillance and maintenance at the four former processing sites.	• DOE continued its active groundwater treatment at two sites under treatment plans that were adopted prior to the initiation of the 2008 5-year plan and continued examining passive treatment options at a third site. DOE also conducted technical studies of groundwater conditions at the sites. • DOE continued long-term surveillance and maintenance at the four sites.	DOE (NRC, Navajo Uranium Mill Tailings Remedial Action program, NNEPA)	Met
Highway 160 site	• Understand the best approach for characterization of the site and any required cleanup based on that characterization.	• DOE and NNEPA completed characterization and cleanup at the site, which included disposing of 5,700 cubic yards of waste at DOE's disposal cell in Grand Junction, Colorado. NNEPA also began addressing contaminated structures located near the site.	DOE (NNEPA)	Met
Tuba City Dump	• Work together with the Navajo Nation and Hopi Tribe to assess whether interim actions are needed to address urgent threats; complete a remedial investigation and feasibility study (RI/FS) and select a remedial action; complete cleanup at the site by December 2012.	• The Bureau of Indian Affairs (BIA), in consultation with the tribes, found no imminent threat to drinking water wells; BIA installed a fence around the site, implemented quarterly groundwater monitoring, and conducted additional studies. BIA and its contractor implemented the agency's settlement agreement with EPA by conducting various assessments of contamination at the site but did not complete the RI/FS. Also, DOE funded a study to characterize the groundwater between the Rare Metals processing site and the landfill. EPA did not select a remedial action for the site, and BIA did not complete cleanup at the site.	BIA (EPA, NNEPA, Hopi Water Resources Program, DOE)	Not met
Health conditions	• Continue to diagnose and treat known health	• IHS continued to diagnose and treat health conditions. IHS also initiated a program that holds community forums	IHS (CDC)	Met

Table 2. (Continued)

2008 5-year plan objective	Targets established by federal agencies in 2008	Key actions taken by federal agencies from 2008 through 2012	Lead federal agency (federal and tribal agency partners)	GAO assessment
	conditions in eligible Indians. • Support a university-led Navajo uranium assessment and kidney health project. • Develop plans for future health studies, and assessment and surveillance of health conditions.	to enhance education about uranium-related health impacts; provides health screenings in communities; updates patients' health records to include information about exposures to uranium; and provides uranium-specific training to health care workers. • IHS staff completed medical evaluations on kidney health project participants living in one region of the Navajo reservation. • An epidemiologist on detail to IHS from CDC helped evaluate options for future health studies and developed plans for assessment and surveillance of health conditions.		

Sources: GAO analysis of information provided by BIA, DOE, EPA, IHS, and NRC.

Note: In addition to establishing targets under eight objectives, the 2008 5-year plan included a more detailed discussion of each objective which, in some cases, identified specific actions the federal agencies planned to take and interim milestones the agencies planned to meet. This table shows the targets established in the 5-year plan but does not identify all actions contained in the plan.

[a] Initial assessments involve field surveys of each mine, including on-the-ground radiation scans at all accessible mines. As described here, detailed assessments evaluate the extent to which people are likely to be exposed to radiation at each mine in order to determine whether mines are eligible for listing on the National Priorities List.

[b] EPA also identified three contaminated structures for which the agency did not provide financial compensation or rebuild because they were rebuilt by the owner or had been abandoned. In July 2012, EPA awarded a $3 million grant to Navajo Community Housing and Infrastructure Department to rebuild nine houses in Haystack, New Mexico. The Navajo agency accepted the grant in February 2013, and work on rebuilding the houses was under way during our site visit in July 2013.

[c] Department of Housing and Urban Development's Community Development Block Grants were also used to fund these projects.

Further, ATSDR worked with the Navajo Nation, the University of New Mexico, and IHS to develop and begin the Navajo Birth Cohort Study, a health study that is intended to improve the understanding of the relationship between uranium exposures and human health—specifically that of mothers and babies—on the Navajo reservation. According to the study proposal, for the Navajo Nation, congenital anomalies remain the leading cause of infant deaths, and the infant mortality rate among the Navajo people is 8.5 deaths per 1,000 live births, compared with 6.9 deaths per 1,000 live births overall in the United States. ATSDR awarded a research cooperative agreement to the university in August 2010, and ATSDR and the university received approval to begin recruiting participants in February 2013; this approval occurred after a lengthy review process that included obtaining multiple, separate approvals, including from the university, the Navajo Nation, and the Office of Management and Budget (OMB).[33] The Navajo Nation and others expressed frustration about the length of time spent developing and approving the study, which took longer than anticipated for a variety of reasons. For example, one reason for the overall amount of time is that OMB did not approve the information collection necessary for the study, or take other actions, within the 60-day regulatory deadline, but rather approved it after more than 300 days.[34] Now that the study is under way, however, ATSDR officials told us that it has already had positive outcomes in Navajo communities. For example, according to ATSDR officials, recent observations of increased levels of prenatal care across the reservation may be a result of the outreach and community education that has occurred as part of the study. Moreover, according to ATSDR's study proposal, the results of the study will answer long-standing questions on whether or not exposures to uranium wastes and other environmental contaminants are associated with adverse birth outcomes or developmental delays on the Navajo reservation. Navajo Nation officials, however, stated that the Navajo Birth Cohort Study is just a small step and that more comprehensive studies are needed to better assess the health effects of uranium contamination on the Navajo people.

	Radiation levels measured at mines				
Distance from potentially inhabited home or structure	At or above 10x background	Between 2x and 10x background	Below 2x background	Unknown[a]	Total[b]
Within 1/4 mile	37 (highest priority) / 2[c]	6[d] (highest priority) / 27	10	-	82
Farther than 1/4 mile	188	154	61	-	403
Total	227	187	71	36	521

■ Highest priority mines

Source: GAO analysis of EPA data.

Note: Background radiation is the radiation that is emitted from naturally occurring materials, such as from radium that is present in nearly all soils and rocks, or from other elements found in the earth's crust, including thorium and potassium. The Environmental Protection Agency (EPA) considered radiation levels of more than 2-times background levels as evidence of an observed hazardous release that may require further investigation under the Comprehensive Environmental Response, Compensation, and Liability Act (CERCLA).

[a] EPA officials said that radiation levels are unknown at these mines because they could not access the mines to gather radiation data, either because of locked gates or fences or remote, mountainous locations.

[b] Figures in this column do not add to 521 since the mines with unknown radiation levels are not included in the Total column.

[c] There are 2 mines with radiation levels at or above 10-times background and within one-quarter mile of a potentially inhabited structure that are not included on the list of 43 highest priority mines. According to an EPA Region 9 official familiar with the mines, these mines are not included on the list because EPA did not have radiation data for them at the time EPA and the Navajo Nation Environmental Protection Agency created the list, and the agencies may add these mines to the list when they update it in the future.

[d] Of these 6 mines, 5 are within 200 feet of a potentially inhabited structure and 1 is located near the Northeast Church Rock mine, which is among the 43 highest priority mines.

Figure 6. 521 Abandoned Uranium Mines on or Near the Navajo Reservation, Radiation Levels, Distance From Homes or Structures, and the 43 Highest Priority Mines.

We found that some of the agencies' actions during the 2008 5-year plan period yielded additional benefits. For example, outreach to affected communities was an important component of some of the objectives under the 2008 5-year plan, although the plan did not include a strategy for coordinating agencies' outreach. Regardless, federal agencies began to coordinate these efforts, and, for example, held five joint workshops for stakeholders, including members of Navajo communities affected by uranium contamination, during which the agencies presented information about their efforts and solicited feedback. Federal agencies also partnered with Navajo agencies on some outreach efforts, which was important for their success in some cases. For example, EPA and NNEPA officials told us EPA relied heavily on NNEPA's outreach staff to communicate with affected community members in identifying and addressing contaminated structures. NNEPA outreach staff's ability to speak Navajo and their familiarity with Navajo cultural practices allowed them to work more effectively with community members than if EPA had conducted outreach on its own. Other benefits from the agencies' actions included tribal capacity building and career development and education opportunities for Navajos. For example, EPA helped enhance capacity building within NNEPA by training some of its staff to assess potentially contaminated structures, and it also provided job training to 20 Navajo hazardous waste workers through EPA's Superfund Job Training Initiative program. In another example, DOE continued to sponsor a summer internship program to give assistance to American Indian college students—including Navajo students—who are pursuing degrees in science, engineering, and technology.

Agencies Met the Targets in Six Objectives Primarily Because of Additional Federal Resources or Because Targets Continued Previously- Established Efforts

We found that a key reason why agencies met the targets for five objectives in the 2008 5-year plan was because additional resources, mostly federal but also private, were dedicated to their efforts. For their work on the objectives addressing contaminated houses, abandoned mines, unregulated drinking water sources, the Highway 160 site, and treatment of health conditions, the agencies either dedicated more funds and staff resources than during the previous 5-year period or received additional appropriations for work related to Navajo uranium contamination. DOE was able to meet the targets for a sixth objective, regarding former uranium processing sites, primarily because its targets largely represented a continuation of previously required activities.

Additional Resources

In accomplishing the targets under five of the objectives outlined in the 2008 5-year plan, according to agency officials, agencies benefited from dedicating additional resources from their existing budgets, receiving additional appropriations to conduct the work, or leveraging funds from private parties. Examples are as follows:

- *Additional funding and staff time from agencies' existing budgets.* EPA prioritized its work under three objectives of the 2008 5-year plan by dedicating additional resources from its existing budget for addressing contaminated houses, assessing abandoned uranium mines, and addressing unregulated drinking water sources. EPA provided from $1.8 million to $7.8 million annually to the Region 9 Superfund Removal program to fund the program's Navajo uranium work during the 5-year plan period—a significant increase over the previous 5-year period. For example, from fiscal years 2008 through 2012, EPA reported that it expended $22 million on efforts to identify and address contaminated houses and other structures, compared with the $1.5 million it expended on similar efforts in the preceding 5 years. Throughout the 2008 5-year plan period, the additional Superfund Removal program funds allowed EPA Region 9 to increase the amount of money it spent on the Navajo work even as the national Superfund budget decreased, according to a senior EPA Region 9 official. Further, EPA officials told us that they conducted work that went beyond the 5-year plan targets because of the increased funding the agency dedicated to Navajo uranium work. Specifically, these officials said they could not have completed the removal action at the Skyline mine without the increased funding since the Region 9 Superfund removal program's prior budget would have been insufficient, and there was no potentially responsible party to contribute funds. In addition to increased funding, EPA Region 9 also increased the number of full-time equivalent employees that it dedicated to its Navajo uranium work from approximately 3.68 in 2008 to 6.95 in 2012. Similarly, IHS reported that its Navajo Area identified nearly $1 million from within its existing budget that it used to support the creation of a uranium-related health screening program, which was established in 2010.

- *Additional funding from the American Recovery and Reinvestment Act of 2009 (Recovery Act).*[35] EPA and IHS used Recovery Act funds for some of the water infrastructure projects that were selected to serve Navajo communities in which contaminated, unregulated water sources had been identified. For example, in fiscal year 2009, EPA contributed $3 million in Drinking Water Infrastructure Grants Tribal Set Aside funds, and IHS contributed about $2 million from the IHS Recovery Act Sanitation Facilities Construction Fund toward a nearly $10 million, 50-mile extension of a water main to the communities around Sweetwater, Arizona. An EPA official familiar with the project told us that it will supply water to homes within the vicinity of four contaminated, unregulated wells, including a well that had the highest uranium levels of all unregulated water sources tested during the 5-year plan period.
- *Additional appropriations.* In fiscal year 2009, DOE received a $5 million appropriation to carry out a remedial action of the Highway 160 site.[36] The 2008 5-year plan included a target for assessing the site and identifying the best path forward, but not for completing cleanup at the site. According to DOE and NNEPA officials involved with the project, having the resources available to fund assessment and cleanup work allowed the agencies to move forward and complete the cleanup more quickly than they had anticipated. Moreover, NNEPA and DOE also used the appropriated funds to begin to address recently discovered, contaminated structures in the area of the Highway 160 site and the nearby Rare Metals processing site.[37] In addition, ATSDR officials we spoke with said they would not have been able to fund the Navajo Birth Cohort Study without additional appropriations for such research. These officials said ATSDR received an increase of $2 million in funding for fiscal

year 2010 to begin the study, and that the agency has subsequently put that amount toward the study.[38]

- *Leveraging funding from private, potentially responsible parties and other federal agencies.* According to EPA officials, the agency was able to complete some of the work that went beyond the targets in the 2008 5-year plan, including conducting the interim time-critical removal actions at the Quivira mine and others, because of funding that came from private, potentially responsible parties. Specifically, EPA issued an administrative order to one of the former operators of the Quivira mine to conduct and pay for the interim removal action. In addition, EPA and NNEPA used funds from a bankruptcy settlement with another potentially responsible party to pay for the interim actions at three other mines or mine-related sites.[39] Without funds from the private, potentially responsible parties, EPA officials said they would not have been able to conduct these actions during the 5-year plan period. Further, EPA officials said that funds from the bankruptcy settlement were instrumental in providing the initial funding for the agency's efforts to pursue potentially responsible parties at other abandoned uranium mines on the Navajo reservation.
- Federal agencies' ability to share resources was also an important factor in meeting the targets in at least one objective. Specifically, IHS officials told us the agency would not have been able to contribute funding to all 13 drinking water infrastructure projects funded during the 5-year plan period without combining its funds with funds from EPA and the Department of Housing and Urban Development. HIS officials said the agency's ability to fund these drinking water projects would have been limited because IHS's share of some of the projects' costs would have exceeded the agency's limit for economic feasibility. For example, if the agency had to solely fund a $4.75 million project in Dennehotso, in the northern part of the reservation, it would have cost the agency about $44,000 per home, an amount that would have been considered economically infeasible for IHS in fiscal year 2009, the year the project was funded. However, contributions of $2 million from EPA and $1 million from the Department of Housing and Urban Development reduced IHS's per-home cost so that the agency was able to participate in funding the project that provided piped drinking water to 107 homes that did not previously have piped water.

Overall, the federal agencies reported spending $121 million on work performed under the 2008 5-year plan.[40] These amounts do not include the approximately $17 million in private funds spent during the 5-year plan period, including by potentially responsible parties at the Northeast Church Rock mine, the Quivira mine, and from the bankruptcy settlement, according to the federal agencies' January 2013 summary report. In contrast, agencies reported spending approximately $42 million during the prior 5 years, and more than half of that amount was spent by DOE at the four processing sites. Figure 7 compares the amount of funds spent by the federal agencies under each objective in the 5-year plan period with the previous 5 years. Because the 2008 5-year plan did not include an overall cost estimate for conducting the work, we cannot determine whether the total amount spent by the agencies was in keeping with their expected costs. The 5-year plan included estimated costs of varying specificity for the first 2 years of the plan—2008 and 2009—since agency budgets for the first 2 years were already in place at the time of the plan's development but not for the final 3 years of the plan.

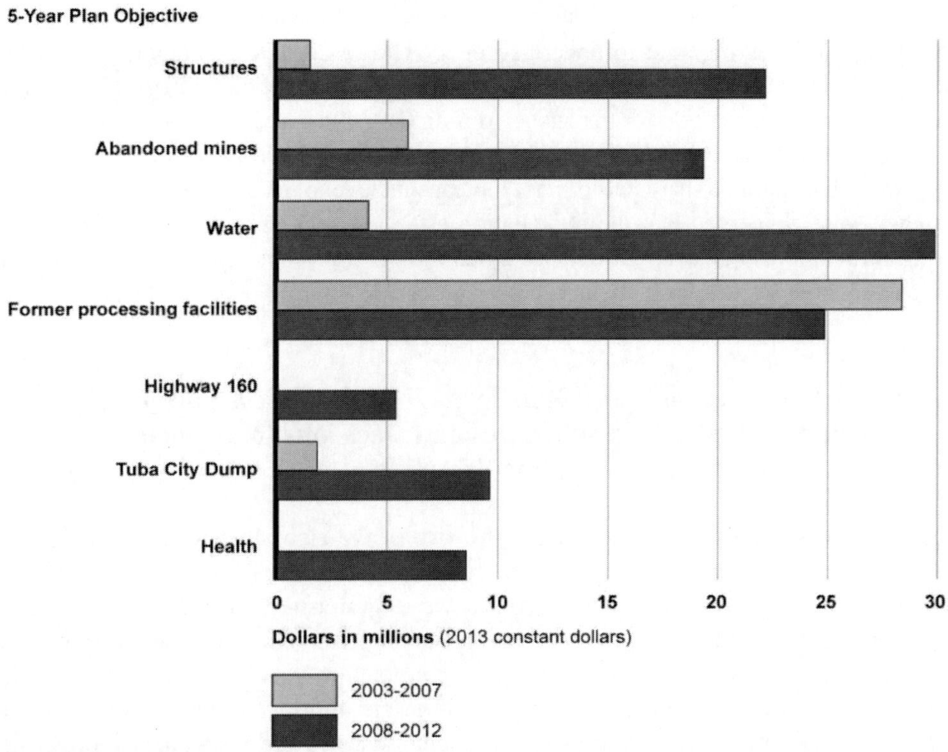

Source: GAO analysis of Agencies' data.

Notes:

The Agency for Toxic Substances and Disease Registry, Bureau of Indian Affairs (BIA), Environmental Protection Agency (EPA), and Indian Health Service reported obligations, and the Centers for Disease Control and Prevention and Department of Energy reported direct outlays. The figure includes only the agencies' direct costs and does not include staff time. The Nuclear Regulatory Commission (NRC) is not included in the figure because it did not report any direct costs, although NRC staff reported working on activities related to Navajo uranium-related sites, including the former processing facilities and the Northeast Church Rock mine.

According to EPA officials, EPA's Northeast Church Rock mine expenditures are included in the Abandoned mines totals for 2003-2007, but most of these expenditures have been reimbursed by the private, potentially responsible party.

BIA reported obligations of its 2-year appropriations but did not report the year the obligation was made. With input from BIA officials, we assigned the obligations to one fiscal year to create a data set that is comparable to the other agencies'.

Dollars do not include $1.35 million in General Assistance Program grants provided to the Navajo Nation by EPA's Office of Water from 2008 to 2012 for uranium-related work.

Figure 7. Federal Agency Expenditures on Actions Taken to Address Uranium Contamination on or Near the Navajo Reservation, 2003-2007 and 2008-2012.

Targets That Continued Previously Required Efforts

For one objective of the 2008 5-year plan—addressing contamination at former uranium processing sites—DOE set targets that largely continued previously authorized activities that the agency was already undertaking, which according to DOE officials, helped the agency accomplish the targets. DOE set targets to continue to address groundwater contamination at three of the sites and to continue long-term surveillance and maintenance at all four sites—actions the agency was required to undertake under UMTRCA. DOE officials told us that continuing to carry out the already approved groundwater remediation strategies at the sites was the most appropriate action for the agency during the 2008 5-year plan period since those were the actions they were explicitly authorized to conduct. Navajo Nation officials told us that they were disappointed that DOE did not increase its level of effort at the sites. They also told us they were concerned that the remediation efforts that DOE is implementing are not achieving sufficient results, and that it appears that the agency is not expected to complete its efforts to treat contaminated groundwater in the foreseeable future.

Agencies Did Not Meet the Targets in Two Objectives for a Variety of Reasons, Including Optimistic Schedules and Decisions to Perform Additional Work That Extended Time Frames

Federal agencies did not meet the targets for two of the eight objectives in the 2008 5-year plan—cleanup of the Northeast Church Rock mine and the Tuba City Dump—for a variety of reasons, including that the schedules were optimistic and ambitious, and EPA decided to increase outreach work at the Northeast Church Rock mine and assessment work at the Tuba City Dump before identifying final cleanup actions for the sites. EPA and BIA officials told us they estimated the schedules based on the information they had at the time, but neither agency anticipated the need for additional steps in the assessment process and therefore did not include these steps in their schedules. Officials from both agencies said they deliberately created ambitious schedules for these sites, in part, to acknowledge the threats they posed and to make it clear that the agencies were committed to cleaning them up. Work remains for both agencies to complete the cleanups at the two sites, and the agencies expect that time frames will likely extend beyond the agencies' 2014 5-year plan and that federal costs will be in the tens of millions of dollars at each site.

Northeast Church Rock Mine

EPA did not meet its target to complete cleanup of the Northeast Church Rock mine in part because its schedule was optimistic and ambitious. According to the 2008 5-year plan, EPA expected to select the removal action for the site in December 2008; however, this selection did not occur until September 2011. According to EPA officials familiar with the project, selecting the removal action took longer than anticipated for a number of reasons. First, completing the cleanup assessment for the site took 8 months longer than planned. Second, after EPA issued the cleanup assessment, the agency postponed selecting the removal action by 2 years so that agency officials could better understand and attempt to address community concerns. Over this 2-year period, EPA conducted 10 public meetings and brought in a Navajo peacemaker to facilitate discussions and improve communication between the agency and the community. According to EPA officials we spoke with, in order to further respond to community concerns, EPA also began work on some predesign analyses that are normally conducted at a later stage in the cleanup process. EPA officials told us they felt the meetings were valuable and that they have conducted more outreach at this site than at most other sites, but community members we spoke with said they remain frustrated with the decision process and disappointed with the outcome.

Third, when estimating the schedule under the 2008 5-year plan, EPA Region 9 officials did not anticipate that additional approval processes would be necessary to implement the removal action for the site that EPA ultimately selected. EPA's selected action involves disposing of approximately 1 million cubic yards of mine waste within an existing disposal cell for mill tailings at a former uranium processing site. The site is located less than 1 mile from the Northeast Church Rock mine and is regulated and managed by NRC and EPA Region 6, respectively. The former operator of the processing site—which is also the former operator and a potentially responsible party for the mine—currently holds a license from NRC for the existing disposal cell at the former processing site. According to EPA Region 9 officials, for EPA to transfer waste from the mine to the disposal cell, EPA headquarters officials determined that EPA Region 6 would need to approve a Record of Decision, which took 18 months to complete.

A number of steps remain for EPA to fully meet the target of cleaning up the mine. As of February 2014, EPA was in the removal predesign phase of the cleanup process. NRC and DOE are both participating in an EPA- led design work group since NRC will transfer responsibility for the site to DOE once the processing site is closed for long-term surveillance and maintenance pursuant to UMTRCA. Once the design phase is complete, the former operator of the processing site must submit a license amendment request and receive an amended license from NRC before disposing of the mine waste at the former processing site. This former operator and potentially responsible party for the mine is expected to implement the removal action if and when NRC issues the amended license. In addition, NNEPA officials told us they have concerns regarding groundwater contamination at the site that have yet to be examined.

EPA's current schedule estimate is to complete the removal action in 2020. EPA officials, however, acknowledged that this schedule is also optimistic since it assumes that NRC's approval process for the license amendment will take 1 year, and that the cleanup itself will take 4 years. NRC officials also said they felt the schedule was too optimistic, and they told us that NRC's safety and environmental reviews will take approximately 2 years but, if a public hearing on the license amendment is requested, the approval process could take up to 5 years.[41] An EPA project manager for the mine told us EPA is working with NRC to revise the schedule to provide 2 years for the license amendment approval process in order to better account for NRC's process. Moreover, although the former operator and potentially responsible party at Northeast Church Rock mine is taking the lead for the cleanup, the government will pay up to 33 percent of future cleanup costs; in 2009, EPA estimated that these total future costs could be $44 million.[42]

Tuba City Dump

BIA also did not meet its targets in the 2008 5-year plan for the Tuba City Dump, in part, because the schedules were optimistic and ambitious. Under the plan, BIA was to (1) complete a set of studies to assess whether interim actions were warranted to protect Hopi water supplies, including drinking water wells, by mid-2008; (2) create a work plan for, then conduct and complete a RI/FS by late 2009; and (3) complete a remedial action by the end of 2012. Partway through BIA's implementation of the 5-year plan, in August 2010, BIA entered into a settlement agreement with EPA to conduct the RI/FS. Under the settlement agreement, BIA's work is subject to EPA's approval, and EPA will select the remedial action. As of the end of the 5-year plan period in 2012, BIA had completed the interim action studies, which found the dump did not pose an immediate threat to the wells, and it had implemented certain actions recommended by the studies, including installing a fence around the perimeter and conducting a detailed analysis of one location with high levels of uranium. BIA developed the RI/FS work plan but had not completed the plan's required work. EPA had not selected a remedial action and, therefore, BIA had not begun or completed a remedial action.

Further, BIA's actions under the 5-year plan took longer than expected, for various reasons, which also contributed to the agency not meeting the targets. First, BIA spent nearly 1 year longer than expected conducting the interim action studies, and implementing the recommended actions took an additional year that had not been accounted for in the 5-year plan. BIA officials said they underestimated the amount of time needed to complete these efforts. Second, BIA spent more time developing the work plan for the RI/FS than had been anticipated, in part because EPA directed it to significantly revise the work plan. Under the settlement agreement, BIA was responsible for submitting a work plan for EPA approval that specified the activities and deliverables, as well as deadlines, for BIA in the development of the RI/FS.[43] The work plan and its deadlines are legally enforceable once EPA approves the work plan. EPA approved the initial work plan developed by BIA in January 2011, and BIA issued a $2 million task award for its implementation in June 2011.[44] However, in July 2011, 1 month later, EPA notified BIA that it would need to revise the RI/FS work plan, which significantly increased the amount of work to be performed. BIA then spent an additional year working with EPA revising the work plan, which EPA ultimately approved in July 2012. According to EPA correspondence to BIA, although it would delay completion of the RI/FS, the additional investigative work was necessary to resolve conflicting interpretations of data collected over the previous years of assessments and to support a defensible selection of a remedial action for the site. EPA officials said the additional work has yielded valuable information, including determining whether groundwater contamination at the site can reach nearby Hopi drinking water wells.[45] Hopi tribal leaders, however, told us that although they appreciate the additional understanding that has been gained through the RI/FS, they are frustrated that the federal agencies have continued to dedicate resources to conducting additional assessments instead of cleanup actions,[46] especially in light of the fact that, as of 2013, BIA had overseen assessment work at the site for more than 10 years.

Third, implementing the RI/FS has taken longer than expected by BIA under the 2008 5-year plan, in part because BIA conducted additional work under the work plan at EPA's direction. In addition to the work EPA directed BIA to add in 2011, EPA subsequently required BIA to conduct additional field investigations. EPA officials explained that the scope of an RI/FS is often changed in response to conditions found on the ground, and that the Tuba City RI/FS has been typical in that respect. According to EPA and BIA documents, conducting this additional fieldwork contributed to BIA missing some of the work plan deadlines.

Moreover, project and contract management challenges faced by BIA have also contributed to the length of time spent on the RI/FS. BIA officials told us they had communication problems with the agency's contractor for the RI/FS and performance problems regarding the quality and timeliness of the contractor's deliverables. For example, in correspondence with the contractor, BIA noted multiple instances when the contractor was late in providing draft deliverables to BIA, which did not provide sufficient time for BIA to review the deliverables before they were due to EPA. In addition, when BIA completed its review of the deliverables, it found they did not all meet the terms and conditions of the contract. BIA also noted in its correspondence with the contractor that the contractor's performance problems began soon after the contract was signed. After the problems continued to mount, according to BIA officials, they worked informally through phone calls and e-mails to correct the performance problems; however, BIA did not formally notify the contractor of the problems and require corrective action until about 16 months after the problems began.[47] During that time period, BIA modified the contract four times, each time increasing the work to be performed in accordance with direction from EPA; these modifications totaled nearly $1.6 million, about an 80 percent increase above the value of the original contract. By adding work to the contract without correcting the contractor's poor performance and adding stronger performance provisions, BIA was effectively rewarding the contractor for its poor performance. In hindsight, BIA officials responsible for managing the contract told us, had they known the problems would not improve, they would have initiated formal action against the contractor sooner; however, they were reluctant to further delay the project. Had they terminated the contractor for default, BIA would have had to award a new contract, taking a minimum of 90 days, plus the additional time it would take to bring a new contractor up to speed to perform the contract. The BIA officials said, instead, they prioritized meeting the deadlines in the work plan and avoiding the delay of awarding a new contract. These officials told us that the RI/FS contract, initially valued at approximately $2 million, is not typical for their region and is much larger than any other contract they manage. For example, the next largest environmental contract in BIA's Western region is worth $300,000. Learning from its challenges in managing the RI/FS contract will become even more important to BIA in the next few years as the agency moves from assessment to cleanup work after a remedial action is selected. At that time, BIA officials said their agency will award and manage a new contract, one that is even larger and more complicated that will increase costs significantly. In August 2011, we reported that incorporating lessons learned from past contracts is an important element of successful acquisition planning when preparing to award a new contract.[48] Through this process, agencies ensure that knowledge gained from prior acquisitions is used to refine requirements and acquisition strategies. Without examining lessons learned from managing the RI/FS contract and considering these lessons as part of the acquisition planning process for the remedial action contract, BIA could face contract management challenges on a larger scale.

Further, according to EPA and BIA officials, BIA's management of the project also contributed to BIA's missing some legally enforceable deadlines in the work plan within months of EPA approving it in July 2012. Specifically, BIA did not comply with the settlement agreement's terms for requesting an extension to these deadlines in the work plan.[49] As a result, BIA was subject to stipulated penalties under the settlement agreement for the deadlines it missed.[50] EPA officials told us the agency held the penalties in abeyance; as a result, EPA did not calculate the total amount of the penalties.[51] As EPA noted in correspondence to BIA, the missed deadlines only led to a few weeks of direct delays to the work plan schedule, but the missed deadlines used much of the contingency, or slack, in the schedule, meaning any future delays could not be absorbed without directly lengthening the project.[52] In 2013, according to EPA officials, BIA notified EPA that it was going to miss another work plan deadline, however, BIA again did not submit the extension request before the deadline passed, potentially subjecting it to additional stipulated penalties.[53]

Further complicating its management of the project, we found that the schedule BIA used to manage its responsibilities under the RI/FS was not created based on best practices for effective scheduling. We have reported that a sound schedule is comprehensive, well-constructed, credible, and controlled.[54] The RI/FS schedule generated by BIA's contractor and approved by EPA minimally met these criteria. For example, we could not verify that the schedule included all the actions needed to complete the RI/FS, which is an essential practice in ensuring that the schedule is comprehensive. If a project schedule does not fully reflect the scope of the project, it can result in unreliable estimated completion dates and delays. In another example, neither BIA nor EPA regularly updated the schedule based on actual progress, an important aspect of a controlled schedule. BIA officials explained that they do not keep a copy of the schedule file that they can update; BIA relies on its contractor to update the schedule, and EPA maintains control of the master schedule file for the RI/FS. Without an updatable version of the schedule, BIA cannot effectively monitor its contractor's progress and cannot evaluate the quality of changes to the schedule proposed by the contractor, which BIA then proposes to EPA for approval. According to BIA officials, not having information about the basis for the proposed schedule changes contributed to the agency proposing a new RI/FS schedule to EPA in 2012 that contained errors and was not achievable. Appendix II contains additional details about our analysis.

A number of steps remain for BIA to meet the target of completing cleanup at the Tuba City Dump. As of February 2014, the full scope of remaining cleanup work—and an estimate of when it may be completed— had not been determined since the RI/FS was ongoing. BIA requested two extensions to the deadlines in the work plan in 2013;[55] as a result, the current deadline for completion of the RI/FS is May 2014, more than 4 years after the completion date in the 2008 5-year plan. According to BIA officials, the May 2014 deadline may not be achieved. For example, BIA officials said they are expecting the schedule to change to allow for additional time for stakeholders' review of a key draft document and for additional analysis requested by EPA in December 2013. In another example, BIA has continued to experience performance problems with its contractor related to timeliness and product quality. These performance problems prompted BIA to send a second formal notification to take corrective action to its contractor in January 2014. Nevertheless, in the short-term, EPA officials said they plan to conduct extensive outreach with local communities as they evaluate the remedial options for the site. Hopi officials we spoke with stressed the tribe's concern over protecting their water sources in the area and told us that having a contaminated dump located on their land is affecting their ability to expand economic development. Because of these concerns, Hopi officials stated that the only acceptable solution is to remove the contamination from the site. DOE and EPA officials involved at the site told us, however, that the data collected thus far indicate the Hopi drinking water wells will not be affected by the dump, and there are other factors limiting development in the area, including the region's scarcity of water.

Based on two potential remedial actions for the site identified by BIA, the agency has estimated that the range of probable future cleanup costs is from $22 million to $72 million.[56] BIA created this estimate range in order to contribute to the Department of the Interior's (Interior) environmental and disposal liability estimate, which is included in Interior's annual financial statement, but we found the estimate was not generated according to the government and industry cost-estimating best practices identified in our 2009 cost estimating and assessment guide.[57] According to BIA officials, the estimate was created according to

Interior's guidance. Nevertheless, according to best practices, cost estimates should be comprehensive, well-documented, accurate, and credible, which are the four characteristics of a high-quality cost estimate of any type, and BIA's estimate does not fully reflect these characteristics. For example, the estimate did not completely define the program, an important aspect of a comprehensive schedule. In response to our questions about some aspects of the work scope that were included in the estimate, including whether future groundwater treatment was included, BIA officials stated that such treatment should be included in the estimate. However, after checking with the contractor that created the estimate, one BIA official involved with managing the project discovered that costs for groundwater treatment were not included in the estimate. Without fully accounting for all future costs, management will have difficulty successfully planning program resource requirements. In response, BIA officials said that they directed the contractor to include these costs in a revised estimate. These officials also said they did not apply all of the best practices when developing the estimate, in part, because it would not have been appropriate for BIA to expend significant resources developing a detailed cost estimate since they have not completed the RI/FS, and that once EPA selects a final remedial action, BIA will work to apply cost estimating protocols when it develops a more detailed cost estimate for the site. Appendix III provides additional details of the results of our analysis of the cost estimate.

Agencies Have Not Estimated the Full Scope of Work, Time Frames, or Costs Needed to Address Uranium Contamination but Recognize That Significant Work Remains

The agencies that implemented the 2008 5-year plan have not identified the full scope of remaining work, time frames, or costs of fully addressing uranium contamination on or near the Navajo reservation, especially at abandoned uranium mines, but have recognized that significant work remains for addressing such contamination beyond the targets in the plan. As a result, decision makers and stakeholders do not have sufficient information about the overall remaining work, time frames, and costs to assess the overall pace of the cleanup efforts. Given that significant work remains to address contamination on or near the Navajo reservation, it is likely that it will take many decades and cost at least hundreds of millions of dollars in additional funding to make significant progress in this area.

Agencies Have Not Identified the Full Scope of Remaining Work, Time Frames, or Costs Needed to Fully Address Uranium Contamination, Especially at Abandoned Mines

Federal agencies that implemented the 2008 5-year plan have not identified the full scope of work needed to fully address uranium contamination on or near the Navajo reservation, especially regarding the abandoned uranium mines and, as a result, they have not estimated total time frames or costs. Prior to the development of the 2008 5-year plan, the House Committee on Oversight and Government Reform reviewed short-term draft action plans that the agencies had prepared. In a January 2008 letter from the House Committee on Oversight and Government Reform to EPA requesting the development of a 5-year plan, the committee wrote the following:

> The draft action plans do not clearly delineate a course of action for fully resolving the problem. Given the extent of the contamination that is already known, it is obvious that the contamination cannot and will not be cleaned up in the 3- to 9-month timeframes covered by the draft plans. We need a 5-year plan from each agency that sets out specific cleanup objectives, specific timeframes for achieving those objectives, and the new authorities and funding, if any, necessary to achieve those objectives. These plans will provide the Congress, the Navajo Nation, and the public with concrete benchmarks against which to measure the progress of the federal agencies in cleaning up the contamination.

In its critique of the short-term action plans, the committee requested of the agencies additional information to understand the full scope of the cleanup effort. The same critique is also generally applicable to the 2008 5-year plan because it too did not contain information on the full scope of the cleanup and instead provided targets for achieving incremental progress under the plan. For example, it is still unclear what percentage of the overall cleanup effort was expected to be achieved in the 2008 5-year plan or how many additional 5-year plans may be necessary to fully address the contamination. As we discussed above, the agencies stated that the 2008 5-year plan focused on addressing over the 5-year period what they identified as the most urgent uranium-related problems and was not intended to be a long-term plan for dealing with the entirety of the contamination. EPA officials involved with coordinating the development of the agencies' 2014 5-year plan told us this plan also will not include the full scope of the cleanup work.

EPA officials cited a variety of reasons for not having identified estimates of the full scope, time frames, or costs of cleanup, including at the abandoned mines. These officials explained that providing such high- level, general estimates of required work, time frames, or costs is not consistent with how EPA cleans up contaminated sites under CERCLA. The agency typically develops detailed, site-specific information on a site- by-site basis, and then estimates costs and schedules based on that specific information. They said the agency generally does not create even rough estimates if cleanup actions have not been selected or if they do not know the total number of mines that will need cleanup. These officials also said that a number of other uncertainties remain. More specifically,

- *Incomplete information about the extent of contamination.* According to the January 2013 report summarizing the agencies' accomplishments under the 2008 5-year plan, EPA and NNEPA's actions resulted in an improved understanding of the scope of uranium contamination at the mines on the reservation, and the agencies identified and prioritized 43 mines that pose the highest risk to surrounding communities. EPA, however, does not know the full scope of cleanup actions that will be necessary to address these highest priority mines (see app. IV for more information about the status of each of the 43 mines), and EPA officials said they expect that some number of the rest of the 521 abandoned mines will also need cleanup, but they do not know what that number will be. EPA officials said that they need additional information about, for example, the location and volume of waste present at each site before they can identify the scope of cleanup actions.[58] However, EPA officials told us they have begun making assumptions about what work may be needed at the highest priority mines based on the site-specific information they have already collected. These officials stated that they expect that most of the highest priority mines will need removal actions, involving excavating and disposing from a few thousand to hundreds of thousands of cubic yards of mine waste at each mine, and a few of these mines may warrant longer-term remedial actions where surface water and/or groundwater may be contaminated.

- *Uncertainty about potentially responsible parties.* According to EPA officials, the total number of abandoned mines that will have a potentially responsible party to lead or contribute funding for assessment and cleanup work is unknown, and this number will affect the scope of work, time frames, and costs necessary to clean up abandoned mines using federal funds. As of February 2014, EPA had signed agreements with potentially responsible parties regarding 24 mines and received money from a bankruptcy settlement for use at another 49 mines—these actions covered 9 of the 43 highest priority mines. EPA Region 9 officials said they are continuing to pursue potentially responsible parties, but the total number of mines that could ultimately be subject to agreements with such parties may be limited, in part, because of difficulties associated with identifying parties more than 50 years after mines were abandoned. Further, there are other reasons why the government's ultimate share of the cleanup costs is unknown. If the federal government is a potentially responsible party at a site, it is liable for the cleanup costs even if a viable nonfederal potentially responsible party is also identified. Also, in November 2013, the Navajo Nation formally stated its intent to file a claim against the United States, and DOE in particular, for reimbursement of its cleanup costs at the abandoned mines on the reservation if a cooperative approach is not successful, which could further affect the government's share of those costs.
- *Uncertainty about disposal options.* Another uncertainty that affects the scope, time frames, and costs of the remaining abandoned mine work is where the mine waste will ultimately be disposed. Currently, the Navajo Nation's position is that all remaining contaminated materials from uranium mines and at processing sites should be excavated and disposed off of Navajo lands. As a result, it is unclear where the volumes of mine waste will be disposed. As of January 2014, the Navajo Nation was working on drafting legislation to create a Uranium Commission under Navajo law that is expected to recommend options for mine waste disposal. However, according to a Navajo Nation official involved with the process, this commission is not expected to make any disposal recommendations until sometime in the next few years.

Even when significant uncertainties regarding the scope of work and available funding remain, however, we have reported that agencies can create high-level estimates of costs and time frames that can be useful for decision makers and stakeholders.[59] For example, EPA can base these estimates on the information it currently has regarding the removal actions that may be necessary at most of the highest priority mines. Specifically, according to our 2009 cost estimating and assessment guide of government and industry cost-estimating best practices, agencies can create high-level cost estimates—for example, rough order of magnitude estimates—for efforts, even with significant uncertainties, that can inform decision makers as they evaluate resource requirements. These cost estimates are often in the form of a range to correspond with the level of uncertainty associated with the estimate and can be developed in short time frames of weeks or months. Although not budget-quality estimates, these types of estimates can be used in planning and can be created before detailed requirements are known. Typically, according to our 2009 cost estimating and assessment guide, an estimate should be revised and contain more detail as the agency obtains more site-specific information and the effort becomes better defined, and the estimate should become more certain as actual costs begin to replace earlier estimates. For example, as EPA obtains more detailed information about the site-specific characteristics at each of the highest priority mines, it would be able to update the scope of its estimate, bringing more certainty. According to our 2009 cost estimating and assessment guide and our 2012 schedule assessment guide,[60] agencies can also create high-level schedules that are linked to cost estimates, based on stated assumptions, and that incorporate uncertainties regarding future activities through a schedule risk analysis. The risk analysis provides agencies with a range of dates that correspond with levels of confidence in the ability to meet those dates.

As further evidence that it is possible to develop these types of high-level estimates, the National Defense Authorization Act for Fiscal Year 2013 requires the Secretary of Energy, in consultation with the EPA Administrator and the Secretary of the Interior, to undertake a review of and prepare a report on abandoned uranium mines across the United States that previously provided uranium ore for the nation's nuclear defense activities.[61] According to DOE documents, the agency plans to issue a report in July 2014 that will include information about the potential costs and feasibility of reclaiming or remediating abandoned uranium mines, including the mines on or near Navajo lands. According to a DOE presentation on the draft report, the report is expected to contain cost estimate ranges based on the amount of uranium ore produced at the mines, among other assumptions. A DOE official involved with developing the draft report told us that the cost estimate ranges are not specific to mines on the Navajo reservation but are based on production size categories of mines across the United States that provided ore to the Atomic Energy Commission. This work by DOE could be a good starting point for a high-level cost estimate to clean up the uranium mines on or near the Navajo reservation; however, based on the statutory requirements for the study, we neither anticipate that it will provide information on the full scope or costs of any other activities covered in the 2008 5-year plan nor any time frames.

Although EPA, DOE, BIA, IHS, and NRC provided some information on high-priority cleanup issues in their 2008 5-year plan, the agencies did not provide the House Committee on Oversight and Government Reform with overall estimates of the remaining scope of work, time frames, and costs of fully addressing uranium contamination on or near the Navajo reservation as requested. Without an estimate of the remaining scope of work, time frames, and costs to fully address uranium contamination, especially at the abandoned mines, decision makers and stakeholders neither have the information they need to assess the overall pace of the cleanup efforts, nor do they have a basis to put the agencies' accomplishments under the 2008 5-year plan into perspective. Navajo Nation officials and other stakeholders told us that they want the federal agencies to describe the full scope of work that remains to fully address the contamination.

Agencies Recognize That Significant Work Remains to Address Uranium Contamination

Although the agencies have not identified the full scope of work that remains to address uranium contamination on or near the Navajo reservation, through implementing the 2008 5-year plan, federal and tribal agencies have compiled information that shows that significant work is needed. For some plan objectives, the agencies have developed a significant long-term scope of work, including the objectives of providing regulated, piped drinking water to Navajo residents in uranium-affected areas, treating groundwater contamination at the former processing sites, and at the abandoned mines. For example, to help lower the number of Navajo residents without access to regulated, piped drinking water in their homes and to continue reducing the use of unregulated and potentially contaminated water sources, IHS developed a list of 145 potential water infrastructure projects that would serve approximately 3,300 homes that do not have piped water.[62] IHS, however, considers just 36 of the 145 projects—serving about 1,000 homes—as economically feasible to fund, according to IHS documents, so it is unclear how many of the 145 projects will ultimately be undertaken by IHS. In another example, according to DOE officials, the agency will continue its active groundwater remediation work at the Rare Metals and Shiprock processing sites, but the future scope of work at the sites is unclear.

This is, in part, because the remediation systems that were designed to address contamination from millions of gallons of water contaminated by mill tailings that entered the ground at these sites are not performing as anticipated.[63] As a result, DOE has not made as much progress toward meeting water quality standards as it originally projected. According to DOE officials, they plan to revise the two sites' groundwater compliance action plans beginning in 2014 and 2015, and these revised plans will dictate the future scope of work at these sites.

For EPA's work at the abandoned uranium mines, although many uncertainties remain about the full scope of work needed to clean up the mines, EPA and Navajo Nation officials said that they recognize that the amount of work will be significant. During the 2014 5-year plan period, EPA officials said, in order to obtain additional information needed to select removal or remedial actions at the highest priority mines, the agency plans to conduct additional assessments at 41 of the 43 highest priority mines, beyond the initial screening information gathered during the 2008 5-year plan period, in cooperation with potentially responsible parties where applicable. EPA and NNEPA officials said these additional assessments range from, at a minimum, scanning the entire site to identify the likely boundaries of contamination and conducting tests to estimate the volume of waste present, to conducting more thorough assessments.[64] These officials told us they do not anticipate conducting cleanups at any mines without potentially responsible parties that would require full funding from EPA during this time period. Appendix V contains additional information about the future work associated with the other 2008 5-year plan objectives not discussed here.

Full Scope of Additional Work Will Likely Take Many Decades to Complete

Given what is known about the significant scope of work that agencies have recognized as remaining to address uranium contamination on or near the Navajo reservation, it is clear that, at current funding levels, it could take at least many decades to complete.

For example, at the abandoned uranium mines, EPA officials said that they are assuming that most of the remaining 42 of the 43 highest priority abandoned mines will need additional removal actions. For one of the highest priority mines—the Skyline mine—EPA has conducted a removal action to clean it up, but the waste remains on the reservation. In the absence of an EPA-estimated time frame, we roughly estimated time frames using information from EPA officials about the number of removal actions that they said they assumed EPA will need to fund and the costs of the agency's removal action for the Skyline mine.[65] Specifically, assuming viable nonfederal potentially responsible parties can be identified for about half of these highest priority mines, which EPA officials said is a reasonable early assumption, federal funds would be necessary to cover the full cost of removal actions at the other half of these mines, or 21 mines.[66] Over the 2008 5-year plan period (i.e., 2008 through 2012), EPA funded the removal action at the Skyline mine from Region 9's Superfund removal budget, spending about $7 million. Assuming Region 9's Superfund removal budget funding levels from the 2008 5-year plan period continued, it would take EPA 105 years to fund the removal actions at 21 of the highest priority mines. According to our rough estimate, it would take even longer to also address the unknown number of mines without potentially responsible parties that will also need cleanup, but which have not been identified.

Moreover, during the decades-long time frames for conducting cleanup at the highest priority abandoned mines, the Navajo people living near these mines could continue to be exposed to elevated radiation levels that pose a high risk, either by visiting the abandoned mines or by inhaling or ingesting contaminated dust that migrates from the mines into communities. For example, as of February 2014, 38 of the 43 highest priority mines remained physically accessible and/or there were no signs communicating the radiation dangers present at the sites. We visited one such mine—the A&B #3 mine near Cameron, Arizona—in July 2013, where EPA measured radiation levels that were 37-times above background (see figure 8). The mine is not signed or fenced and is located within one-quarter mile of nearby homes. According to a local government official and a Navajo agency official, they have seen evidence that people visit the mine site; they told us they have found children's toys at the site and pointed out vehicle tracks.

In addition, the time frames associated with addressing contamination under other plan objectives are lengthy as well. For example, IHS officials estimated that it would take approximately 38 years to complete 36 of the 145 projects it identified as necessary to provide regulated, piped drinking water to residents of areas affected by historic uranium mining, assuming IHS were the sole contributor of funds and based on IHS's Navajo Area water program budget for fiscal year 2013. During this time frame, residents may continue to be exposed to harmful constituents potentially found in unregulated drinking water sources. EPA and NNEPA officials told us that although the official position of the Navajo Nation is that unregulated water sources are not fit for human consumption, they continue to receive reports that Navajo residents use these sources, in part if there is no alternative, but for other reasons as well. In addition to health dangers posed by drinking uranium-contaminated water, which ATSDR and others have linked to kidney disease, IHS and EPA consider the general lack of regulated drinking water to be a health risk because contaminants often found in unregulated sources, such as *E. coli* bacteria, can pose an immediate health danger to people that consume them.[67] Appendix V contains additional information about the potential future time frames associated with the other 2008 5-year plan objectives not discussed here.

The Scope of Additional Work Will Most Likely Cost Hundreds of Millions of Dollars

For the significant amount of remaining work to address uranium contamination that federal agencies recognize is needed, it appears that associated costs could exceed hundreds of millions of dollars. For example, in the absence of a cost estimate from EPA for work at the abandoned mines, our rough estimate based on previously incurred costs by EPA at the Skyline mine indicates that EPA's costs to fund removal actions at just half of the highest priority mines, or 21 mines, could be a minimum of about $150 million.[68] This is a conservative, low-end estimate for a number of reasons but, most importantly, because it does not include costs to transport and dispose of waste off-site.[69] According to EPA officials, this is one of the most significant factors influencing cleanup costs; disposing of waste offsite is consistent with the Navajo Nation position that such waste be removed from Navajo lands.

Other federal agencies have developed cost estimates for addressing contamination under other plan objectives, and the costs for these efforts appear to reach into the hundreds of millions of dollars as well. For example, IHS officials estimated that the 36 economically feasible drinking water projects would cost about $35 million, and that all 145 projects would cost about $195 million to complete.[70] In another example, DOE has estimated that its various actions at the former processing sites will cost about $193 million over the next 75 years, but DOE officials said that this estimate will need revision based on new groundwater remediation plans at the Rare Metals and Shiprock sites.[71] Appendix V contains additional information about the potential future costs associated with the other 2008 5-year plan objectives not discussed here.

View of the A&B #3 mine.

Source: EPA.

EPA measuring elevated gamma radiation on a rock at the A&B #3 mine.

Looking at nearby homes and structures from the A&B #3 mine.

Figure 8. Photographs of A&B #3 Mine, Uranium-Bearing Rock Located On-site, and Nearby Homes and Structures.

AGENCIES FACE CHALLENGES MEETING FUNDING NEEDS AND ENGAGING AFFECTED COMMUNITIES, THOUGH OPPORTUNITIES EXIST FOR IMPROVING RELATIONSHIPS WITH THE NAVAJO AND HOPI PEOPLE

Federal agencies face a variety of challenges in continuing to address uranium contamination on and near the Navajo reservation, including securing adequate funding and effectively engaging tribal communities. However, federal and Navajo agency officials and community members we spoke with identified opportunities for improving relationships with the Navajo and Hopi people, which could help the federal agencies more effectively engage these communities. In addition, other opportunities exist to enhance collaboration with federal agencies and with Navajo agencies.

Adequate Funding for Continued Efforts to Address Uranium Contamination Is a Key Challenge

One key challenge that federal agencies face is difficulty meeting funding needs with available federal resources. Specifically, according to EPA Region 9 officials, funding for EPA's efforts to assess and clean up abandoned uranium mines under its Superfund removal program, especially those without viable private, potentially responsible parties, and to provide clean and safe drinking water is expected to decrease from funding levels that have been available from 2008 through 2012. EPA officials told us that reducing the human health risks associated with abandoned uranium mines on the Navajo reservation is a priority and that the agency intends to continue providing funding as resources allow. However, these officials stated that declining Superfund removal program resources nationally will likely result in a reduction in funding available to conduct removal actions at Navajo abandoned mines from the level available in previous years. An EPA Region 9 official familiar with funding tribal drinking water projects on the Navajo reservation also told us that federal resources available for drinking water infrastructure projects are expected to continue to decrease.[72] In addition, as federal funding resources for abandoned mines and water projects become more constrained, such projects on the Navajo reservation may be less likely to receive federal funds because the programs by which they are funded prioritize projects based on risk. For example, under the EPA Superfund removal program, which EPA has used to pay for assessment and cleanup at the Navajo abandoned mines, projects that address emergency situations, such as toxic waste spills, are prioritized for funding over projects that generally are not considered emergencies, like the abandoned mine projects. EPA officials have prioritized the Navajo abandoned mine projects, but these officials told us that there is already a high demand for Region 9's Superfund removal program funds with more projects that warrant selection than available funding.

Moreover, with few exceptions, federal agencies currently have limited options for other sources of federal funding for uranium-related work on the Navajo reservation. For example, according to EPA Region 9 officials, one possible source of funding for the typically longer-term remedial actions that may be warranted at a few of the highest priority mines without potentially responsible parties is the Superfund Trust Fund. Although the Superfund Trust

Fund can be used for removal actions at all sites, it can only be used for remedial actions at sites that are included on the NPL.[73] None of the Navajo abandoned mines are currently listed on the NPL, and, according to EPA Region 9 officials, only a few of the highest priority mines on the reservation may qualify for listing. In general, according to these officials, to score high enough in the Hazard Ranking System to be included on the NPL, an abandoned uranium mine would need to impact a sufficient number of people using a drinking water supply contaminated by the mine, expose a sufficient population to uranium through air or soil contamination, or impact a sensitive environment such as a wetland. However, given the locations and characteristics of the mines on the reservation and the population or environment affected, these officials said they believe most mines do not meet these criteria. For example, EPA officials told us that even the Navajo mines located near communities affect relatively small populations given the dispersed nature of the Navajo population, and the surrounding desert conditions mean that most of the mines do not appear to impact surface water. EPA officials said that community members in rural communities, including Navajo communities, have expressed a deep frustration with the ranking system used to determine sites' inclusion on the NPL because they feel that the system unfairly discriminates against small communities. Nevertheless, EPA officials said they will continue to pursue including some of the mines on the NPL in order to use the Superfund Trust Fund to pay for remedial actions.

Another potential source of federal funding is Interior's Central Hazardous Materials Fund (Fund), an appropriation available to pay for Interior's CERCLA response actions. BIA received $162,000 from the Fund for the Tuba City Dump site, in 2008 and 2009. According to BIA officials, this money was used, in part, to search for another potentially responsible party at the site, pay for project oversight, and hire a technical consultant. In 2011, Interior issued a memorandum stating that CERCLA response actions on Indian trust lands were no longer eligible to receive money from the Fund. BIA officials told us, based on this memorandum, they believed that they were no longer eligible to obtain money for assessment or cleanup work at the Tuba City Dump. These officials said that they have not requested funding from the Fund to pay for assessment work and do not plan to request funding for the eventual remedial action at the site. However, Interior's memorandum also stated that the Fund would continue to fund cleanup-related activities on Indian trust lands if BIA had received funding for cleanup-related activities, including those undergoing CERCLA assessments, in the past. According to a senior Fund official, because the Tuba City Dump site received Fund money prior to 2011, it is eligible to receive additional funds, including funding for the remedial action work. Pursuing this funding is important for two reasons: (1) once the remedial action has been selected, BIA's funding requirements are likely to increase substantially (BIA estimated the remedial action could cost about 3- to 10-times as much as the RI/FS), and (2) federal standards for internal control encourage agencies to strive for efficiency in their use of resources.[74] Since the Tuba City Dump is located in BIA's Western region, the agency has paid for its work at the site out of that region's budget. According to BIA officials, BIA has prioritized the Tuba City Dump project over other projects in the region that also need funding, which has resulted in some projects not receiving funding when BIA's costs at the Tuba City site were especially high.

In addition, EPA is pursuing two other funding sources to contribute to the work at the abandoned mines. First, EPA and a potentially responsible party have reached an agreement to settle a pending lawsuit which, if approved by the judge presiding over the lawsuit, could result in approximately $1 billion for the agency's and the Navajo Nation's cleanup efforts at 50 mines and other contaminated sites.[75] Second, EPA has sought to involve DOE in assessments and potentially cleanups at mines that do not have potentially responsible parties. In June 2013, EPA Region 9 corresponded with DOE, stating that DOE's financial assistance with developing and implementing an approach to conducting assessments, interim actions, and cleanups at highest priority mines on the Navajo reservation is essential. Senior DOE officials told us they interpreted the letter as encouraging DOE to play a larger role in addressing the contamination from the mines on the reservation by funding assessment and cleanup at some of the mines. In January 2014, DOE responded to EPA that, although the National Defense Authorization Act for Fiscal Year 2013 required a DOE report on abandoned uranium mines, DOE was not given budget authority in fiscal year 2013 to remediate uranium mines, and its authority to take remedial actions under UMTRCA, which was limited to the former uranium processing sites and vicinity properties, has expired.

Engaging Tribal Communities Continues to be a Key Challenge

Another key challenge faced by federal agencies is identifying ways of more effectively engaging with tribal communities. According to outreach plans prepared by DOE and EPA, and other documents prepared by various federal agencies, engaging with communities is important for a number of reasons, including soliciting feedback on the decision-making process, obtaining meaningful input into cleanup decisions, and working with community members to determine how best to limit exposures to uranium contamination. During the 2008 5-year plan period, federal agency officials increasingly recognized the importance of community engagement and began building bridges in the communities where they conducted work, both by developing relationships themselves, and by funding Navajo agency officials' outreach work. Even with the federal agencies' increased attention to outreach activities, agency officials and community members we spoke with said that the need for increased and improved outreach is great. Nevertheless, federal agencies face challenges in their ongoing efforts to effectively engage Navajo communities for at least four reasons: (1) building trust may require significant time and effort on the Navajo reservation, (2) the number of outreach staff is small compared with the size of the reservation, (3) commonly used tools for engaging communities may not be effective in Navajo communities, and (4) federal agencies have not coordinated their outreach efforts.

Building Trust

Agency officials and community members we spoke with said that although building trust among the Navajo people is necessary to effectively engage local communities, it will take significant time and effort. One reason for this is that many members of the Navajo community distrust outsiders—especially those representing the federal government—because of historical events, related to both uranium mining and the government's treatment of Native people. The federal government's inconsistent attention to uranium-related issues on Navajo lands in recent decades may also have contributed to a lack of trust among community members. For example, according to EPA Region 9 officials, EPA compiled a list of potentially contaminated houses on the Navajo reservation in the 1970s, but it did not take steps to ensure that all houses on the list were assessed for contamination until it began work under the 2008 5-year plan.[76] Distrust of the federal government is also exacerbated by concerns about ongoing issues, such as fears that federal agencies will issue permits for new uranium mining near the Navajo reservation before contamination from previous mining is fully addressed.[77]

Another reason why building trust will be a challenge is because Navajo community members are concerned that the federal agencies that worked on the 2008 5-year plan may not have a long-term commitment to addressing uranium contamination, according to Navajo Nation officials and some stakeholders. For example, Navajo community members have expressed disappointment that the 2008 plan encompassed just 5 years' worth of work when, in their view, fully addressing the effects of contamination will take decades of commitment. A long-term commitment—along with completing cleanup work—could help build trust with Navajo communities. One challenge the federal agencies will continue to face in addressing these concerns, however, is that, in some cases, the agencies are limited in the types of long-term commitments that they can make. For example, EPA officials explained that the agency cannot commit to cleaning up even the 43 highest priority mines at this time because they do not have dedicated funding for addressing the highest priority mines that do not have potentially responsible parties and are not listed on the NPL. This is in contrast to the situation at the former uranium processing sites, where DOE must prepare and implement a long-term surveillance plan for disposal sites, in accordance with NRC regulations implementing UMTRCA.[78]

In addition, building trust is a challenge because Navajo Nation officials and some stakeholders told us they are frustrated by what they see as examples of environmental injustice.[79] These are instances when uranium contamination on the Navajo reservation appears to be treated differently than contamination on non-Indian lands, such as the community in Moab, Utah, where DOE is excavating a large mill tailings pile and disposing of it elsewhere.[80] In another example of environmental injustice on the Navajo reservation, high-level Navajo officials and others have said that the release of radioactive materials from the uranium processing site near Church Rock, New Mexico, has received far less attention nationally than the radioactive release at Three Mile Island—which occurred 4 months earlier—although the amount of radioactive materials released in the later incident was significantly greater.

Finally, some tribal agency officials we spoke with told us they believe the federal agencies had fostered mistrust by sometimes overstating their progress in addressing uranium contamination on the Navajo reservation. Specifically, some high-level Navajo Nation officials have stated that they believe the federal agencies have understated the scope of the uranium contamination problem on the Navajo reservation and have overstated the federal agencies' efforts in addressing the problem. For example, Navajo Nation officials said they were frustrated that the federal agencies' January 2013 summary report, published at the end of the 2008 5-year plan period, highlighted the agencies' accomplishments but did not identify or communicate the larger context: that, overall, significant progress has not been made in addressing uranium contamination on the reservation.

Limited Number of Outreach Staff

Federal and tribal agency officials we spoke with said that the number of federal and tribal agency outreach staff working on engaging Navajo communities about uranium-related issues is very small compared with the size of and conditions on the reservation. Outreach staff are responsible for engaging communities that are spread out over three states across the 24,000 square-mile reservation. Many of these communities are not only remote, but they are also difficult to access because of harsh terrain and rough roads.

- *Dedicated outreach staff.* As of November 2013, EPA Region 9 had the full-time equivalent of 1.5 outreach staff working on Navajo uranium issues, including only one staff member who speaks Navajo. EPA outreach staff are not based on the Navajo reservation, and travel time between the reservation and Region 9 headquarters in San Francisco limits the amount of time that staff are able to spend engaging communities. In addition to EPA outreach staff, NNEPA has one outreach staff member dedicated to uranium issues, and one additional staff member who incorporates outreach in her work. NNEPA's dedicated staff person, however, is responsible for activities in addition to community outreach, such as interfacing with the media on uranium-related issues, and this staff person told us that she cannot meet all the outreach needs put before her. For example, she said that coordinating with IHS on outreach events could be a full-time position, but that she is limited by other demands on her time.
- *Other staff conducting outreach.* Other federal and tribal agencies that worked on the 2008 5-year plan conduct outreach as part of their activities, but they do not have dedicated staff working full-time on uranium-related outreach to Navajo communities. For example, IHS has two staff members who engage with communities on uranium issues, in addition to their other responsibilities, and DOE has one staff member who performs outreach activities in addition to her site management responsibilities. The Navajo Abandoned Mine Lands Reclamation/UMTRA department's public affairs staff, under the UMTRA program's cooperative agreement with DOE, has engaged Navajo communities to discuss concerns about former processing sites located on the reservation. In addition, the Navajo Nation Division of Health has used ATSDR funding to hire staff who conduct outreach activities as part of their responsibilities related to implementing the Navajo Birth Cohort Study.

Limitations of Common Outreach Tools

For a variety of reasons, commonly used tools for communicating with communities—such as disseminating written materials, including brochures or e-mails, and putting up signs and fencing off contaminated areas—may generally be less effective on the Navajo reservation. For example, written materials may be less effective because Navajo and Hopi are traditionally spoken languages—not written languages—and many community members learned English as a second language. Also, many residents are not connected by Internet or telephone in their homes. According to EPA and NNEPA officials, the most effective way to communicate with many members of Navajo communities is through face-to-face interactions, which requires trusted native speakers and is more time-consuming than written communications. Furthermore, although signs and fences may be used to communicate information about risks from contamination, they may be less effective on the Navajo reservation. In part because of differences in how Navajo people traditionally view the land, it is generally not acceptable to restrict the use of reservation lands, although there are some exceptions, such as those related to grazing uses and home sites. As a result, knowledgeable

EPA officials told us they did not believe that signs and fences would be sufficient to limit access to contaminated areas because they felt signs and fences would be disregarded. As an example, these officials told us that a mining company had erected a fence to restrict access to an area contaminated by uranium, but, rather than staying out of the area, a community member had instead used the fence to contain his livestock, confining them in the very area to which the agency was trying to restrict access. Furthermore, according to these officials, physical structures such as signs and fences are difficult to maintain in remote areas of the reservation, where vandalism and theft pose challenges.

Moreover, providing information, regardless of delivery method, may itself be a limited tool for changing behavior because, in many cases, no acceptable alternatives are available. For example, according to Navajo agency officials and community members with whom we spoke, some community members—because they do not have a better alternative— continue to get their drinking water from unregulated livestock wells that may be contaminated with uranium and other toxins, even though some of these community members understand that doing so is unsafe. Some community members we spoke with said that they used unregulated water sources for domestic purposes, such as cooking and drinking, and that more education would not be effective in changing this behavior until better alternatives were made available.

Limited Federal Agency Collaboration on Outreach Efforts

Federal agencies that worked on the 2008 5-year plan have not generally coordinated their outreach efforts. Although the agencies began hosting joint workshops for stakeholders in 2008, according to agency officials, the agencies generally have conducted their own public meetings in communities without inviting other agencies to participate. Agency officials told us that not coordinating outreach poses a challenge to effectively engaging communities because community members often expect these meetings to cover a variety of uranium-related issues, regardless of whether those issues fell within the jurisdiction of the agency present. For example, according to an EPA official we spoke with, when EPA conducted outreach related to abandoned uranium mines, community members often had questions about other uranium-related topics, such as health effects. In some cases, the limited scope of issues covered in community meetings has caused significant frustration among community members, according to EPA and IHS officials. An EPA official told us that this may hamper the efforts of outreach staff to build relationships in tribal communities. Because the costs of attending community meetings can be high for both federal agencies and community members—many of whom travel significant distances to attend the meetings—EPA officials said they realized it is important to maximize each contact that they have with affected communities. Officials from multiple agencies told us they recognize the value in coordinating on outreach and have begun to coordinate their efforts by, for example, holding joint community meetings. For example, in March 2012, IHS and DOE met jointly with community members to discuss the Shiprock former uranium processing site, including concerns about health impacts. In another example, in 2013, EPA and IHS jointly hosted a health screening in one Navajo community for residents who had been living in contaminated homes that were being demolished through EPA's actions to address and replace contaminated structures.

Opportunities Exist to Improve Federal Agencies' Relationships with the Navajo and Hopi People

Federal and Navajo agency officials and community members we spoke with identified a number of opportunities that federal agencies could pursue to improve relationships with the Navajo people, as well as with the Hopi people affected by the Tuba City Dump. Opportunities identified for improving relationships with tribal communities include the following:

- *Provide information on long-term scope of work.* Federal officials and community members identified opportunities for federal agencies to provide a more complete picture of the scope of the uranium contamination problem and their progress toward addressing the problem. For example, some community members, including participants at one of the agencies' stakeholder workshops, told the agencies they would like to see the next interagency plan cover a period longer than just another 5 years because they believe it is clear that the amount of work remaining will take significantly longer than 5 years. Stakeholders said that including information about the long- term scope of work in the next plan would increase their ability to hold the agencies accountable and provide a benchmark against which they can measure the agencies' progress.
- *Conduct in-person outreach.* Federal officials and community members identified opportunities for federal agencies to improve their relationships with Navajo communities—and the Hopi people affected by the Tuba City Dump—by conducting in-person outreach where possible, although such methods are resource intensive. For example, in a community involvement plan created to guide its outreach related to the Tuba City Dump, EPA noted that distributing information in small group or door-to-door settings assists in developing trust and keeping misunderstandings of new materials and information to a minimum. According to the plan, the Hopi people have regularly requested that federal agencies rely on in-person outreach methods when feasible. The plan stated that although it is not feasible to distribute all information in person, doing so conveys to community members that they are important and are part of the process.
- *Establish agency offices on or near the reservation.* One way agency officials identified to establish a more constant presence on the reservation would be for federal agencies that do not already have offices nearby, including EPA and DOE, to set up and assign technical and outreach staff to offices in the area. This would increase the amount of time that staff can interact with communities, since typically EPA and DOE staff travel to the reservation from California or Colorado. According to a Navajo Nation official we spoke with, having staff on or near the reservation would increase the federal agencies' ability to connect with communities, especially since it would help increase their cultural awareness and sensitivity. One EPA staff member who spent 2 months working on-site during a mine cleanup said that his consistent presence during that time—as well as the extensive outreach he conducted over a longer period—allowed him to build strong relationships with community

members, which in turn increased that community's acceptance of the cleanup remedy.

- *Partner with community organizations.* Some stakeholders we spoke with said that opportunities may exist for federal agencies to more effectively engage Navajo communities by partnering with trusted community organizations. For example, the president of a nonprofit community organization that works to ensure that the Navajo people—especially those affected by uranium contamination in a remote region of the reservation—have access to safe drinking water and economic development, among other things, told us that he would welcome a partnership with federal agencies to conduct outreach on uranium-related issues. Such a partnership would take full advantage of the organization's existing connection with Navajo communities. Representatives from this organization and others told us that community organizations are often a trusted source of information, and their involvement would lend credibility to federal and tribal agencies' engagement efforts.
- *Promote job creation and training.* Navajo officials we spoke with also told us that the federal agencies could help improve relationships by identifying opportunities to promote job creation and training on the Navajo reservation as part of the efforts to address uranium contamination. Officials we spoke with said the federal government should provide more funding for new positions within the tribal agencies to address uranium contamination. They said that, in 2007, the Navajo Nation had identified the need for 20 new full-time employees within NNEPA to address uranium contamination, but that federal agencies awarded funding for just 2 additional employees from 2008 to 2012. Tribal officials also said that they would like to see the federal agencies provide job training programs similar to the one that EPA offered during the 5-year plan period.
- *Issue a formal apology.* Some stakeholders, including Navajo community members, told us they felt that receiving an official apology from the federal government for failing to ensure that the companies conducting uranium mining to support U.S. nuclear weapons development were protective of the environment and public health would go a long way toward improving relationships.[81]

Additional Opportunities Exist for Federal Agencies to Enhance Interagency Collaboration and Collaboration with Navajo Agencies

Agency officials we spoke with said that additional opportunities exist for the federal agencies to enhance both interagency collaboration at the federal level and collaboration with Navajo agencies. Specifically, EPA, IHS, and DOE officials identified a number of opportunities for increased interagency collaboration on efforts to engage tribal communities. For example, an IHS official involved with conducting uranium-related health screenings told us that the joint health screening event conducted with EPA in 2013 was a success and that, in addition to duplicating such efforts in other affected communities, there may be additional opportunities for enhanced interagency collaboration to help ensure that the Navajo people receive health screenings as well as information on how they can most effectively protect themselves from uranium contamination. More specifically, the IHS official told us that there may be opportunities to work with partners, including CDC and EPA, to develop informational videos on the health effects of uranium exposure that could be screened in IHS clinics. In addition, EPA officials told us that they have initiated a pilot effort to provide more coordinated outreach in the Cameron region of the reservation. In that region—where potentially responsible parties will be conducting extensive assessment and some cleanup of abandoned mines in the coming years—EPA and NNEPA plan to work with other partners, including other federal agencies, to provide a more coordinated approach to engaging the communities in discussions about, among other things, steps the federal agencies and community members can take to mitigate exposures to hazardous uranium contamination. Further, EPA officials involved in coordinating the 2014 5-year plan told us that they plan to engage the other federal agencies in developing and including a coordinated outreach strategy in the plan to better ensure that the agencies maximize each contact that they have with affected communities by, for example, providing the communities with information on a variety of uranium-related issues. Officials from the other agencies agreed that they would engage with EPA to develop or support such a strategy. This is consistent with one of the key practices that, in October 2005, we reported can help enhance and sustain collaboration among federal agencies—establishing mutually reinforcing or joint strategies—which can assist partner agencies in aligning their activities and resources, among other things.[82] We have reported on other key practices to enhance and sustain interagency collaboration, including, for example, for collaborating agencies to define and agree on their respective roles and responsibilities.[83] In doing so, collaborating agencies can identify how their collaborative efforts will be led, clarify who will do what, organize their joint and individual efforts, and facilitate decision making.

Federal and Navajo agency officials also identified opportunities for the federal agencies to enhance collaboration with Navajo agencies, some of which also present opportunities to enhance capacity building. For example, an EPA official we spoke with said that the agency could potentially train NNEPA staff to perform the more detailed assessments that EPA has been conducting to determine whether houses are contaminated and warrant replacement. In addition, Navajo Nation officials told us that they would like the federal agencies to work with them to identify as many opportunities as possible for the federal agencies to partner with the Navajo agencies on uranium-related work. These officials pointed to the partnership between DOE and NNEPA at the Highway 160 site—where NNEPA led the implementation of the cleanup work—as a particular success that they would like to see replicated in other areas. EPA Region 9 officials also pointed to the partnership between EPA and the Navajo Community Housing and Infrastructure Department, a Navajo agency that is helping to replace some of the contaminated houses on the reservation.

According to both federal and tribal agency officials, federal agencies could also enhance their collaboration with Navajo agencies by including additional tribal agencies in future efforts. For example, EPA and DOE identified potential opportunities for enhanced collaboration with the Navajo Abandoned Mine Lands Reclamation Program, which was not involved in the abandoned mine work conducted under the 2008 5-year plan, although the UMTRA program under the same department has been working with DOE at the former uranium processing sites. The Navajo Abandoned Mine Lands Reclamation Program was active in abandoned uranium mine-related efforts during the plan period by, among other things, maintaining reclamation work previously conducted to mitigate physical hazards at the mines. According to Navajo Abandoned Mine Lands Reclamation Program officials, they plan to continue this maintenance work at reclaimed mine sites in the future. According to both federal and tribal agency officials we spoke with, EPA could potentially collaborate with that program and NNEPA to help ensure that, where feasible, any additional maintenance work on reclaimed uranium mines is done in coordination with NNEPA and EPA to help further reduce radiological hazards.[84] EPA officials told us that they have begun talking with Navajo Abandoned Mine Lands Reclamation Program staff to identify ways to work together, whereas in the period of the 2008 5-year plan, the program and NNEPA each generally operated independently from one another. According to DOE, EPA and DOE have agreed to invite officials from NNEPA and both programs within the Navajo Abandoned Mine Lands Reclamation/UMTRA Department to participate in activities of either federal agency. In another example, a CDC official told us there may be opportunities for CDC to increase its collaboration with the Navajo Division of Health to improve available data on how cancers impact the Navajo people.

CONCLUSION

From 2008 through 2012, six federal agencies increased their overall efforts to address the legacy of uranium contamination that remained on the Navajo reservation after uranium mining and processing ceased, spending more than $120 million on various actions, including assessments of abandoned mines and cleanups of contaminated homes and other sites. However, nearly 30 years since the last active uranium mine on the Navajo reservation ceased production, federal agencies do not have comprehensive information about the extent of the contamination or the total scope of work—and associated time frames and costs—required to fully address it, especially the contamination found at the abandoned mines. When requesting the 2008 5-year plan, policymakers were looking for a comprehensive course of action for fully resolving the problem of uranium contamination on or near the Navajo reservation. Given that the scope of the 2008 5-year plan focused on addressing the most urgent problems, and the agencies' next 5-year plan is not expected to identify the full scope of work that remains, it is unclear how many 5-year plans at this rate would be needed to estimate the remaining scope of work, time frames, and costs for fully addressing the contamination. While many uncertainties exist, it is possible to generate useful, high-level estimates of the work, time frames, and costs in a short period of time based on the information the federal agencies currently possess. However, absent a statutory requirement to develop such a comprehensive estimate, it appears unlikely that the agencies will undertake such an effort. Without more comprehensive information about the overall remaining scope of work, time frames, and costs needed to address contamination across the reservation, including at the abandoned mines, stakeholders and decision makers do not have a basis on which to assess the overall pace of the cleanup efforts; and without this information cannot put the accomplishments of the 2008 5-year plan—or any future plans—into perspective, and cannot make effective resource allocation decisions.

Effectively engaging with tribal communities is a key challenge facing federal agencies in the efforts to address contamination on Navajo and Hopi lands. The 2008 5-year plan did not contain information about how the federal agencies would coordinate their outreach efforts to these communities. While the agencies began to integrate their outreach activities over the course of the 5-year plan period, community members continued to express frustration with the agencies' efforts. Creating a coordinated outreach strategy is consistent with the key practice of establishing mutually reinforcing or joint strategies that we have reported can help enhance and sustain interagency collaboration and help agencies better align their activities and resources. Such an effort should also identify how the collaborative effort will be led, clarify who will do what, organize their joint and individual efforts, and facilitate decision making.

In addition, assessment work conducted by BIA at the Tuba City Dump has yielded information about the contamination at the site that has provided significant value to decision makers. In doing so, however, BIA has experienced a number of challenges, some concerning contract management. BIA has missed enforceable deadlines, subjecting BIA to stipulated penalties under its settlement agreement with EPA. Moreover, BIA continued to increase the value of the contract while the contractor was not performing according to the contract's terms and conditions. BIA is nearing the end of its management of the current RI/FS contract; these contract management challenges, however, if left unaddressed, will become even more pertinent in the next few years as BIA moves into the cleanup phase after a remedial action is selected. At that point, the agency will award and manage an even larger and more complicated contract that will increase costs significantly. Without examining lessons learned from managing the RI/FS contract and considering these lessons as part of the acquisition planning process for the remedial action contract, BIA could face contract management challenges on a larger scale. Further, BIA did not fully follow best practices in estimating the schedule for the RI/FS, which was not fully comprehensive or controlled. Without control over the schedule, BIA cannot effectively monitor its contractor's progress and cannot evaluate the quality of changes proposed by the contractor. BIA's estimate of probable future costs for the cleanup at the Tuba City Dump also did not always reflect the characteristics of a comprehensive high-quality cost estimate. Without fully accounting for all future costs, management will have difficulty successfully planning program resource requirements. Further, significantly more funds will likely be needed to implement the remedial action that will be selected for the Tuba City Dump site. Given this increased need and other competing interests for BIA's limited resources, other funding sources for remedial actions, such as Interior's Central Hazardous Materials Fund, become more important. The Tuba City Dump site is eligible to receive funds from the Fund for the RI/FS, as well as the selected remedial action, although BIA officials have not applied for such funding and do not plan to do so. Without leveraging the Fund, BIA will have difficulty meeting the funding needs for the remedial action cleanup phase of the project.

MATTER FOR CONGRESSIONAL CONSIDERATION

To develop an estimate of the scope of work remaining to address uranium contamination on or near the Navajo reservation, Congress should consider requiring that the Environmental Protection Agency take the lead and work with the other federal agencies to develop an overall estimate of the remaining scope of the work, time frames, and costs.

Recommendations for Executive Action

We are making the following four recommendations in this report:

To ensure that agencies working on the 2014 5-year plan better align their activities and resources, we recommend that the Administrator of EPA; the Secretaries of Energy, the Interior, and Health and Human Services; and the Chairman of the Nuclear Regulatory Commission, as they develop a coordinated outreach strategy to include in the 2014 5-year plan, take action to incorporate key practices in their collaborative effort, such as defining and agreeing on the agencies' respective roles and responsibilities.

In light of the problems BIA has encountered in managing the cleanup at the Tuba City Dump site, we recommend that the Secretary of the Interior direct the Assistant Secretary for Indian Affairs to take the following three actions:

- identify and examine any lessons learned from managing the RI/FS contract and consider these lessons as part of the acquisition planning process for the remedial action contract,
- employ best practices in creating the schedule and cost estimates for the remedial action cleanup phase, and
- apply for funding from Interior's Central Hazardous Materials Fund in order to help meet the funding needs for the remedial action cleanup phase of the project.

Agency and Third Party Comments and Our Evaluation

We provided a draft of this report for review and comment to the Environmental Protection Agency; the Departments of Energy, the Interior, and Health and Human Services; the Nuclear Regulatory Commission; the Office of Management and Budget; and the Navajo Nation and Hopi Tribe. EPA, DOE, BIA (responding on behalf of Interior), the Department of Health and Human Services, and NRC generally agreed with our recommendations. Each of these agencies also provided technical comments, which we incorporated as appropriate. The Navajo Nation also provided written comments and technical comments, which we incorporated as appropriate. The Office of Management and Budget and the Hopi Tribe did not comment on our report.

DOE was the only agency that commented on our Matter for Congressional Consideration. Specifically, DOE acknowledged the need to identify the remaining scope of work, time frames, and costs of fully addressing uranium contamination on the Navajo reservation. However, the agency expressed concern about the difficulty of quantifying the full scope of work at this time, given the number of uncertainties that remain. We agree that attempting to quantify, in a detailed manner, the full scope of remaining work is not possible at this time because of the uncertainties we describe in this report, as well as those identified by DOE in its comments. However, we believe that the agencies' estimates can be improved and, for the reasons detailed in our report, consider it essential for Congress to have more comprehensive information about the remaining scope of work in order to assess the overall pace of the cleanup and to make informed resource allocation decisions. In addition, DOE commented on (1) our use of the term "Rare Metals" site instead of the "Tuba City Mill Site" or the "Tuba City Former Processing Site" and (2) an observation in the report about the cleanup of the former processing sites on the Navajo reservation versus another site near Moab, Utah. Regarding the comparison of the former processing sites on the Navajo reservation to the Moab site, DOE provided comments on the unique nature of each site. Regarding the Rare Metals site, we clarified our report to indicate that our use of the term "Rare Metals site" was solely to prevent it from being confused with other contaminated sites in the Tuba City area, including BIA's Tuba City Dump site.

In agreeing with our recommendation that the federal agencies incorporate key practices for enhancing and sustaining interagency collaborative efforts into their coordinated outreach strategy, EPA, DOE, and the Department of Health and Human Services provided additional details about the contents of the draft outreach strategy that they intend to include in their 2014 5-year plan. We are encouraged by elements of the draft strategy, which appear to include some of the key practices for enhancing and sustaining interagency collaboration. For example, the agencies plan to leverage resources to fund a shared coordinator who will direct a community outreach network composed of representatives from the federal and tribal agencies.

In its written comments, BIA asserted that our report disproportionately focuses on the agency's management of the Tuba City Dump site, a site that BIA described as comprising a very small part of the overall problem of uranium contamination across the Navajo reservation. Although we agree that cleaning up the Tuba City Dump is but one part of the agencies' broader efforts to address uranium contamination on the Navajo reservation, we believe that our detailed examination of BIA's and EPA's management of the assessment work at the site was appropriate.

As explained in our report, the Tuba City Dump site was one of the two objectives where the federal agencies did not meet the targets in the 2008 5-year plan. In order to identify reasons why BIA did not meet those targets, we analyzed the agency's management of the Tuba City Dump site, including its project and contract management approaches. In addition, BIA commented that it believes the management of the cleanup has been handled responsibly but that circumstances beyond its control have contributed to delays. We acknowledge in this report that circumstances beyond BIA's control contributed to delays in the Tuba City Dump cleanup; however, BIA's own actions also contributed to these delays. For example, BIA had communication problems with its RI/FS contractor and performance problems regarding the quality and timeliness of the contractor's deliverables. These problems and others led BIA to miss several legally enforceable deadlines in the RI/FS work plan and resulted in BIA failing to complete the RI/FS for the site, an objective of the 2008 5-year plan and a work plan requirement. In its comments, BIA appears to fault EPA and the contractor for the delays and associated stipulated penalties, but BIA did not explain how, if at all, EPA's and the contractor's actions would relieve BIA of its legal obligation to meet the work plan deadlines, or of the consequences of failing to do so.

While agreeing with our recommendations, BIA disagreed with our findings regarding the agency's cost and schedule estimating at the Tuba City Dump. These findings were the basis for our recommendation that BIA employ best practices in creating the schedule and cost estimates for the forthcoming CERCLA remedial action at the Tuba City Dump. BIA asserted that its approach to developing the cost and schedule estimates was reasonable, and did not believe that the best practices found in our cost estimating and assessment guide were applicable, at least in part, because the details of the remedial action to be conducted at the site have not been identified. As explained in this report, the best practices in our cost estimating and assessment guide can be used for projects with significant unknowns, such as the Tuba City Dump, and include specific steps for properly taking those unknowns into account. For example, the guide discusses how every cost estimate is uncertain because of the assumptions that must be made about future projections. It also states that because many of the assumptions made at the start of a project turn out to be inaccurate, it is important to assess the risk associated with changes in assumptions. As described in our report, we found that BIA did not assess such risks when creating its cost estimate. In addition, although BIA concurred with our recommendation to use best practices when creating cost and schedule estimates for the forthcoming remedial action, BIA stated that it does not intend to use these best practices when developing future estimates at the site because it does not believe that they are applicable to environmental cleanup projects. To the contrary, these guides offer best practices that are directly relevant to a wide range of government projects, including environmental cleanups, and we have previously assessed such projects using the criteria in the guides.[85] BIA's refusal to apply cost and schedule estimating best practices is troublesome, especially since, as we note in the report, the remedial action at the site will represent a more significant undertaking than the RI/FS. As such, it is important that BIA have reliable cost and schedule estimates in order to effectively manage the project.

In written comments, the Navajo Nation stated that, overall, our report represented a good start toward illuminating the nature and extent of the damage uranium mining and processing has caused to Navajo lands and people. However, the Navajo Nation also noted some instances where it felt the report fell short. For example, the tribe noted that our projection of the need for hundreds of millions of dollars to address the remaining scope of work that has been identified by the agencies was a significant underestimation of the total projected costs for future remediation at all uranium-related sites on Navajo lands. The tribe stated that it believes that these total costs will be in the billions of dollars rather than hundreds of millions of dollars. As noted in our report and recognized by the Navajo Nation in its comments, our estimate related to the cleanup of highest priority abandoned mines represents a conservative, low-end estimate. While we cannot comment on the accuracy of the tribe's characterization of total future costs, the tribe's estimate further illuminates the need for the federal agencies involved to generate a high-level estimate, as discussed in our report, based on their most current information regarding the remaining scope of work. In addition, the Navajo Nation noted the absence of any discussion of potential groundwater contamination in the draft report, especially at the Northeast Church Rock mine. In response, we have made changes to include information about the status of federal agencies' groundwater assessment efforts.

J. Alfredo Gómez
Director, Natural Resources and Environment

APPENDIX I. OBJECTIVES, SCOPE, AND METHODOLOGY

In this report, we examined: (1) the extent to which federal agencies, including the Environmental Protection Agency (EPA), Department of Energy (DOE), the Department of the Interior's Bureau of Indian Affairs (BIA), the Department of Health and Human Services'

Indian Health Service (IHS), and the Nuclear Regulatory Commission (NRC) achieved the targets identified in the 2008 5-year plan, and the reasons why or why not; (2) what is known about the scope of work, time frames, and estimated costs of fully addressing uranium contamination on the Navajo reservation; and (3) the key challenges, if any, faced by federal agencies in completing this work and the opportunities, if any, which may be present to help overcome these challenges.

To determine the extent to which federal agencies achieved the targets identified in the 2008 5-year plan, we compared the agencies' targets as laid out in the 5-year plan with the actions taken by the agencies and their partners over the 5-year plan period from 2008 through 2012. We identified these actions by reviewing key documents, including the summary report issued by the federal agencies in January 2013. We corroborated information in the documents by interviewing relevant federal agency officials and Navajo and Hopi tribal officials from relevant tribal government agencies—the Navajo Nation Department of Justice, Navajo Nation Environmental Protection Agency, the Navajo Nation Division of Natural Resources' Navajo Abandoned Mine Lands Reclamation/Uranium Mill Tailings Remedial Action Department, and the Hopi Tribe Department of Natural Resources' Water Resources Program—and by obtaining additional documentation and visiting relevant sites across the Navajo and Hopi reservations where federal and tribal agencies have been conducting their work. In addition to the five federal agencies listed above, other relevant federal agencies we spoke with included the Centers for Disease Control and Prevention (CDC) and the Agency for Toxic Substances and Disease Registry (ATSDR), the Office of Management and Budget, and the Department of the Interior's Office of Environmental Policy and Compliance, which manages the Central Hazardous Materials Fund, and the Office of Surface Mining Reclamation and Enforcement, which provides funding to the Navajo Abandoned Mine Lands Reclamation Program. In April 2013 and July 2013, we visited key sites, including the Northeast Church Rock, Quivira, and Skyline mines, as well as abandoned uranium mines and mine-related sites near the communities of Cameron, Cove, and Teec Nos Pos, Arizona; and Haystack, and Casamero Lake, New Mexico; the former uranium processing site in Shiprock, New Mexico; the Highway 160 site near Tuba City, Arizona; and the Tuba City Dump, located on both the Hopi and Navajo reservations and near the Hopi Villages of Moenkopi and the Navajo town of Tuba City, Arizona. We selected these sites based on the level of activity that federal and tribal agencies conducted there during the 5-year plan, and in order to see some of the sites that the agencies have identified as needing cleanup work in the near future.

To identify the reasons why the agencies met or did not meet the targets in the 5-year plan, we reviewed agency documents and interviewed federal and tribal agency officials. We also reviewed federal agency expenditure data for the 2008 5-year plan period (fiscal years 2008 through 2012) and compared it with expenditure data from the previous 5 years (fiscal years 2003 through 2007). These data represent obligations or direct outlays by the agencies and represent the agencies' direct costs and did not include intramural costs, such as staff salaries. We received data from ATSDR, BIA, CDC, DOE, EPA, and IHS; NRC did not provide expenditure data because NRC did not incur any direct obligations during the time period, although it did expend resources for staff time. In order to determine costs in constant 2013 dollars, we adjusted the amounts reported to us for inflation by applying the fiscal year chain-weighted gross domestic product price index, with fiscal year 2013 as the base year. To evaluate the reliability of these data and determine their limitations, we reviewed the data obtained from each agency. For each data source, we analyzed related documentation, examined the data to identify obvious errors or inconsistencies, and compared the data we received with other published data sources, where possible. We also interviewed officials from each agency to obtain information on the internal controls of their data systems. On the basis of our evaluation of these sources, we concluded that the expenditure data we collected and analyzed were sufficiently reliable for our purposes.

To identify what is known about the scope of work remaining to fully address uranium contamination on or near the Navajo reservation, we reviewed available documents and interviewed knowledgeable federal agency and tribal officials. To identify what is known about time frames and costs, we reviewed documentation containing schedule and/or cost estimates or general information about time frames or costs, where available. To create estimates of time frames and costs to clean up the highest priority abandoned mines, we gathered information about the costs associated with the Skyline mine cleanup, which was the one cleanup EPA conducted during the 5-year plan period with the agency's funds, the pace of work conducted under the 5-year plan, and the number of mines that would need full funding from EPA. In addition, we analyzed the extent to which the schedule generated by BIA for the remedial investigation and feasibility study at the Tuba City Dump reflected the four general characteristics for sound schedule estimating, as outlined in our schedule assessment guide: comprehensive, well-constructed, credible, and controlled.[86] We selected this schedule to review because it was the most robust of the available schedules, and it represented an entirely federal effort. We also examined the extent to which BIA's estimate of probable future costs for Tuba City Dump reflected the four characteristics of high-quality cost estimates, as outlined in our cost estimating and assessment guide: comprehensive, well-documented, accurate, and credible.[87] We selected this cost estimate to review since the cleanup will be paid for entirely with federal funds, and it represented a distinct cleanup project rather than an ongoing level of effort. In reviewing BIA's schedule and cost estimates, we analyzed supporting documentation submitted by BIA and conducted interviews with BIA and EPA project managers and staff. We shared our cost and schedule guides and the criteria against which we would be evaluating the estimates with BIA staff. We then compared BIA's methods and approaches for preparing the estimates with the best practices contained in the guides. To ascertain the key challenges faced by federal agencies in completing this work and the opportunities that may be present to help overcome these challenges, we reviewed key documents, including the January 2013 summary report, written materials from the federal agencies' Navajo uranium stakeholder workshops, agency reports, and Navajo Nation laws and position papers. We corroborated and supplemented information in the documents by interviewing relevant federal agency officials and Navajo and Hopi tribal officials. We also interviewed knowledgeable stakeholders, including community members living in areas affected by uranium mining or contamination. For example, we worked with the federal and tribal agencies and others to hold meetings in seven affected communities, which were attended by 50 local government officials and community members. We also spoke with other knowledgeable stakeholders, such as university researchers and representatives of nonprofit and community organizations active on Navajo uranium issues. We met with most of these stakeholders during our July 2013 site visits and at the April 2013 Navajo Uranium Stakeholder Workshop held in Gallup, New Mexico, and spoke with other stakeholders by telephone. We identified stakeholders by performing an Internet and literature search for individuals and organizations involved in relevant issues, attending the stakeholder workshop and identifying participating stakeholders, and requesting referrals from agency officials and stakeholders with whom we spoke. The views of the stakeholders we spoke with are not representative of and cannot be generalized to all stakeholders.

We conducted this performance audit from January 2013 to May 2014 in accordance with generally accepted government auditing standards. Those standards require that we plan and perform the audit to obtain sufficient, appropriate evidence to provide a reasonable basis for our findings and conclusions based on our audit objectives. We believe that the evidence obtained provides a reasonable basis for our findings and conclusions based on our audit objectives.

APPENDIX II. DETAILED RESULTS OF GAO ASSESSMENT OF THE SCHEDULE FOR THE TUBA CITY DUMP REMEDIAL INVESTIGATION AND FEASIBILITY STUDY WORK PLAN

Our prior work has identified 10 best practices associated with effective scheduling.[88] These are (1) capturing all activities; (2) sequencing all activities; (3) assigning resources to all activities; (4) establishing the duration of all activities; (5) verifying that the schedule is traceable horizontally and vertically; (6) confirming that the critical path is valid; (7) ensuring reasonable total float; (8) conducting a schedule risk analysis; (9) updating the schedule with actual progress and logic; and (10) maintaining a baseline schedule. These practices are summarized into four characteristics of a reliable schedule—comprehensive, well-constructed, credible, and controlled. We assessed the extent to which the Bureau of Indian Affairs' (BIA) January 2013 schedule for the remedial investigation and feasibility study (RI/FS) at the Tuba City Dump met each of the 10 best practices,[89] and characterized whether the schedule met each of the four characteristics of a reliable schedule. We found that the schedule minimally met each of the four characteristics of a reliable schedule. As a result, we are concerned about the validity of the dates that were forecasted by the schedule, as well as the identification of the critical path. Without an accurate critical path, management cannot focus on the activities that will be most detrimental to the project's key milestones and finish date if they slip. Table 3 provides the detailed results of our analysis.

Table 3. Assessment of the January 2013 Tuba City Dump RI/FS Work Plan Schedule

Schedule characteristic	Explanation	Criterion met[a]	GAO analysis
Comprehensive	Capturing all activities, assigning resources to all activities, and establishing the durations of all activities. • All activities as defined in the project's work breakdown structure are included; • the labor, materials, and overhead needed to do the work and whether those resources will be available when needed; and	Minimally	While our analysis found a good ratio of detail to milestone activities, we could not determine if the schedule reflected the work necessary to conduct the remedial investigation and feasibility study (RI/FS) work plan. Specifically, there was no work breakdown structure documentation submitted, and only 50 percent of the schedule activities had work breakdown structure elements assigned to them. As a result, we could not validate if all work necessary to complete the project is included in the schedule. In addition, there were only two resources used for the project, and they were both overallocated, one by over 5,000 percent.

Schedule characteristic	Explanation	Criterion meta	GAO analysis
	• how long each activity will take, allowing for discrete progress measurement with specific start and finish dates.		Consequently, it appears that those resources will not be available when needed. Further, our analysis found that the use of multiple duration units caused duration conversion issues, which made the schedule's activity durations unreliable.
Well-constructed	Sequencing all activities, confirming that the critical path is valid and ensuring reasonable total float. The schedule should be constructed so that • all activities are logically sequenced with predecessor and successor logic; • limited amounts of unusual or complicated logic techniques are justified in the schedule documentation; • a critical path that determines which activities drive the project's earliest completion date can be identified; and • total float that accurately determines the schedule's flexibility.	Minimally	We found that 17 of the remaining schedule activities (29.3 percent) were missing logic, 4 had hard constraints without justification (about 7 percent), and there were lags and leads used that cause the sequencing to be questionable. Although hard constraints are useful for calculating the amount of float available in the schedule and, therefore, the realism of the required project finish date and available resources during schedule development, they may be abused if they force activities to occur on specific dates that are determined off-line without much regard for the realism of the assumptions necessary to achieve them. Because of these logic issues, the critical path as calculated by the scheduling software was deemed unreliable. Our analysis found that the critical path contained lags, leads, long-duration activities, and activities with hard constraints, which by definition will always appear as critical. We also found that the driving path could not be traced, and from what few driving activities could be found, they differed from the default critical path. Finally, we encountered task inspector errors that suggested rescheduling a task to its resource's next available time. There were differing amounts of total float, on the critical path in particular, that led to questionable schedule flexibility. A possible contributing factor may be the duration units assigned in the schedule that caused conversion issues, as discussed above.
Credible	Verifying that the schedule is traceable horizontally and vertically, conducting a schedule risk analysis. The schedule should identify • the order of events necessary to achieve aggregated products or outcomes;	Minimally	We found the schedule to be vertically traceable, with dates in the detailed schedule mapping to higher-level management correspondence. We also identified schedule documentation that matched the project start and completion dates in the schedule. In contrast, we found the schedule could not be traced horizontally. For example, the schedule did not respond appropriately when
	• varying levels of activities, supporting activities, and subtasks; • key dates that can be used to present status updates to management; • a level of confidence in meeting a project's completion date based on data about risks and opportunities for the project; and • necessary schedule contingency and high-priority risks based on conducting a robust schedule risk analysis.		significant delays were introduced into the network due to the logic issues discussed above. Furthermore, we received no documentation to substantiate that a schedule risk analysis was performed, so the credibility of the dates in the schedule were also in question.

Table 3. (Continued)

Schedule characteristic	Explanation	Criterion met[a]	GAO analysis
Controlled	Updating the schedule with actual progress and logic, maintaining a baseline schedule. The schedule should be • updated periodically by schedulers trained in critical path method scheduling; • statused using actual progress and logic to realistically forecast dates for program activities; • compared against a documented baseline schedule to determine variances from the plan; • accompanied by a corresponding baseline document that explains the overall approach to the project, defines assumptions, and describes unique features of the schedule; and • subject to a configuration management control process.	Minimally	It is unclear if schedule progress is updated monthly because the schedule's status date was outdated. Furthermore, there was no evidence that a formal schedule narrative accompanied the schedule delivery to the government. While it appeared that the Bureau of Indian Affairs (BIA), BIA's contractor, and Environmental Protection Agency (EPA) officials meet to discuss the schedule, it does not appear that the schedule is used to manage the project because EPA did not want to put additional information about BIA and the contractor's effort in the schedule. Our analysis also found that, relative to the assumed status date, 5 activities had start dates in the past. Although we cannot determine when the baseline was set, all 126 activities have baseline start dates. Project documentation also stated that the schedule was baselined, but the baseline date cannot be found in MS Project. Nevertheless, it is commendable that the schedule is baselined, which allows for analysis and monitoring of date variances. On the other hand, there is no evidence that a schedule baseline document was created for this schedule baseline.

Source: GAO analysis of BIA data.

[a] "Met" means the program provided complete evidence that satisfies the entire criterion. "Substantially" means the program provided evidence that satisfies a large portion of the criterion. "Partially" means the program provided evidence that satisfies about half of the criterion. "Minimally" means the program provided evidence that satisfies a small portion the criterion. "Not met" means the program provided no evidence that satisfies any of the criterion.

APPENDIX III. DETAILED RESULTS OF GAO ASSESSMENT OF THE TUBA CITY DUMP 5-YEAR COST ESTIMATE

To analyze the Bureau of Indian Affairs' (BIA) 5-year cost estimate, dated April 2013, for the Tuba City Dump, we determined the extent to which BIA followed the best practices outlined in the GAO Cost Estimating and Assessment Guide.[90] The guide identifies 12 practices that are the basis for effective cost estimation, including cost estimation for annual budget requests. The guide associates these practices with four characteristics: accurate, well-documented, comprehensive, and credible. The Office of Management and Budget endorsed this guidance as being sufficient for meeting most cost-estimating requirements, including for budget formulation. If followed correctly, these practices should result in reliable and valid budgets that (1) can be easily and clearly traced, replicated, and updated; and (2) enable managers to make informed decisions. BIA created this cost estimate to contribute to the Department of the Interior's (Interior) environmental and disposal liability estimate as part of Interior's annual financial statement. Since a remedial action has not been selected for the site, BIA estimated two options, a low-cost option and a high-cost option. In accordance with Interior's guidance, BIA submitted the low-cost option as part of the liability estimate. We assessed the extent to which the Tuba City Dump 5-year cost estimate from April 2013 met each of the four characteristics associated with cost-estimating best practices.[91] As table 4 illustrates, we found that the Tuba City Dump April 2013 cost estimate minimally met each of these four characteristics.

Table 4. Assessment of the Tuba City Dump 5-Year Cost Estimate

Cost characteristic	Explanation	Criterion met[a]	GAO analysis
Comprehensive	The cost estimate includes all life- cycle costs; the cost estimate completely defines the program, reflects the current schedule, and is technically reasonable; the cost estimate work breakdown structure is product-oriented, traceable to the statement of work, and at an appropriate level of detail to ensure that cost elements are neither omitted nor double-counted; and the estimate documents all cost-influencing ground rules and assumptions.	Minimally	The estimate does not include all life-cycle costs, including both government and contractor costs, to clean up, operate, and maintain the Tuba City Dump. This is because a remedial action has not been selected for the site and because the Department of the Interior's (Interior) guidance suggests that agencies only include future costs, rather than life-cycle costs, in the liability estimates.
Well-documented	The documentation should capture in writing the source data used, the reliability of the data, and how the data were normalized; the documentation describes in detail the calculations performed and the estimating methodology used to derive each work breakdown structure element's cost; the documentation describes step by step how the estimate was developed so that a cost analyst unfamiliar with the program could understand what was done and replicate it; the documentation discusses the technical baseline description and the data in the technical baseline are consistent with the cost estimate; and the documentation provides evidence that the cost estimate was reviewed and accepted by management.	Minimally	According to project officials, the estimate was developed in accordance with Interior's *Environmental and Disposal Liabilities Identification, Documentation and Reporting Handbook*. Though the handbook clearly states what should be included in the source data, we found that the Tuba City Dump estimate did not meet all of the criteria. In addition, the documentation supporting the cost estimate is limited. Neither the cost model nor the supporting documentation details the calculations or the estimating methodology by element. The documentation neither captures how the scope of the project was determined, nor does it capture at a detailed level how quantities of materials or labor rates were determined. As a result, a cost analyst unfamiliar with the Tuba City Dump and the estimating process would not be able to recreate the estimate based on the documentation package received.

Table 4. (Continued)

Cost characteristic	Explanation	Criterion met[a]	GAO analysis
Accurate	The cost estimate results are unbiased, not overly conservative or optimistic, and based on an assessment of most likely costs; the estimate has been adjusted properly for inflation; the estimate contains few, if any, minor mistakes; the cost estimate is regularly updated to reflect significant changes in the program, such as when schedules or other assumptions change, and actual costs so that it is always reflecting current status; variances between planned and actual costs are documented, explained, and reviewed; the estimate is based on a historical record of cost estimating and actual experiences from other comparable programs; and the estimating technique for each cost element was used appropriately.	Minimally	The cost estimate does not include a risk and uncertainty analysis. While the estimate included contingency costs, it is unclear what statistical analysis was performed to derive these numbers. The escalation method was not documented in the cost estimate, and the estimate did not cite the source of the inflation rate. An inflation factor of 5 percent was used in the estimate. Mathematically, the calculations are accurate, and no mistakes are apparent. We credit the agency with keeping a documented log of changes to the estimate, although none of these changes are evident in the estimate itself. No costs were labeled as actuals in the documentation provided, and there is no mention of whether variances between the estimate and actual costs were reviewed. It is unclear from the evidence which estimating technique was used for each element and if that method was appropriate.
Credible	The cost estimate included a sensitivity analysis that identified a range of possible costs based on varying major assumptions, parameters, and data inputs; a risk and uncertainty analysis was conducted that quantified the imperfectly understood risks and identified the effects of changing key cost driver assumptions and factors; major cost elements were cross-	Minimally	The cost estimate does not include a sensitivity analysis. The documentation says there are many unknowns associated with this project that may impact the probable costs. To account for these unknowns, a minus 30 percent/plus 50 percent range was provided in addition to the most probable cost. Without documentation that shows how they arrived at the minus 30 percent/plus 50 percent
	checked to see if results are similar; and an independent cost estimate was conducted by a group outside the acquiring organization to determine whether other estimating methods produce similar results.		contingency range, we have no evidence that statistical analysis was performed. In addition, there is no evidence a risk and uncertainty analysis was conducted that quantified the imperfectly understood risks and identified the effects of changing key cost driver assumptions and factors.

Cost characteristic	Explanation	Criterion meta	GAO analysis
			The documentation provides contingency factors, however, there is no documentation supporting how these numbers were derived. In addition, the documentation does not include the likelihood and severity of consequence for each risk, which work breakdown structure elements are affected by each risk, cost risk ranges by work breakdown structure elements, or a risk correlation matrix by work breakdown structure element.

Source: GAO analysis of BIA data.

[a] "Met" means the program provided complete evidence that satisfies the entire criterion. "Substantially" means the program provided evidence that satisfies a large portion of the criterion. "Partially" means the program provided evidence that satisfies about half of the criterion. "Minimally" means the program provided evidence that satisfies a small portion the criterion. "Not met" means the program provided no evidence that satisfies any of the criterion.

APPENDIX IV. THE 43 HIGHEST PRIORITY MINES, THEIR LOCATIONS, AND THE STATUS OF ASSESSMENT AND CLEANUP EFFORTS

Mine name	Region	Potentially responsible party identified	Fence, sign, or other barricade	Initial assessment (radiation scan)	Interim removal action[a]	Detailed assessment initiated[b]	Final removal action selected
Occurrence B	Central			✓			
Black Jack No. 2	Eastern			✓			
Eunice Becenti	Eastern			✓			
Haystack No. 1	Eastern			✓			
Mac No. 1	Eastern			✓			
Mariano Lake	Eastern	✓	✓	✓		✓	
Northeast Church Rock	Eastern	✓	✓	✓	✓	✓	✓
Quivira (NE Church Rock No. 1)	Eastern	✓	✓	✓	✓	✓	
Ruby No. 3	Eastern	✓	✓	✓			
Section 23	Eastern			✓			
Section 24 (Nanabah Vandever)	Eastern			✓			
Section 25	Eastern			✓			

(Continued)

Mine name	Region	Potentially responsible party identified	Fence, sign, or other barricade	Initial assessment (radiation scan)	Interim removal action[a]	Detailed assessment initiated[b]	Final removal action selected
Section 26 (Desidero Group)	Eastern	✓		✓			
Standing Rock	Eastern			✓			
Charles Keith	North Central			✓			
Firelight No. 6	North Central			✓			
Harvey Blackwater No. 3	North Central			✓			
Mitten No. 3	North Central			✓			
Rock Door No. 1	North Central			✓			
Skyline	North Central		✓	✓	✓[c]	n/a	n/a
Alongo Mines	Northern			✓		✓	
Barton 3	Northern			✓			
Black Rock Point Mines	Northern			✓			
Climax Transfer Station	Northern			✓			
Hoskie Henry	Northern			✓			
King Tutt Point	Northern			✓		✓	
Mesa II ½, Mine 4	Northern	✓		✓			
Mesa III Mine	Northern	✓		✓			
Mesa III, Northwest Mine	Northern			✓			
NA-0904	Northern			✓			
NA-0928	Northern			✓			
North Martin	Northern			✓			
Oak 124, Oak 125	Northern			✓			
Plot 3	Northern			✓		✓	
Plot 6	Northern			✓		✓	
Tsosie 1	Northern			✓			
Hoskie Tso No. 1	Southern			✓			
A & B No. 2	Western			✓			
A & B No. 3	Western			✓			
Boyd Tisi No. 2	Western			✓			

Mine name	Region	Potentially responsible party identified	Fence, sign, or other barricade	Initial assessment (radiation scan)	Interim removal action[a]	Detailed assessment initiated[b]	Final removal action selected
Charles Huskon No. 12	Western	✓		✓			
Charles Huskon No. 14	Western	✓		✓			
Juan Horse No. 3	Western			✓			

Source: GAO analysis of Environmental Protection Agency (EPA) data.

[a] Removal actions conducted at these sites were done as time-critical removal actions and are listed as interim because they did not constitute the final cleanup actions at the mines.

[b] Detailed assessments listed here include remedial action assessments (preliminary assessment and/or site investigation) and removal action assessments (engineering evaluation/cost analysis).

[c] The Skyline mine cleanup is included here because the Navajo Nation considers it to be temporary, given its current position regarding removing all mine waste from the reservation. EPA considers the cleanup complete.

APPENDIX V. ADDITIONAL INFORMATION ABOUT THE REMAINING SCOPE OF WORK, TIME FRAMES, AND COSTS TO ADDRESS URANIUM CONTAMINATION ON OR NEAR THE NAVAJO RESERVATION

The following is a discussion of what is known about the remaining scope of work, time frames, and costs to assess and clean up contaminated structures and to assess and treat health conditions and conduct health research.

Assess and Clean Up Contaminated Houses and Other Structures

The Environmental Protection Agency (EPA) and the Navajo Nation Environmental Protection Agency (NNEPA) have not identified a full scope of work because there is no comprehensive source of information regarding the number of houses that may be contaminated; the agencies do not have an end date for this work nor an overall cost estimate.

Scope of work: EPA Region 9 officials told us they believe living in contaminated homes continues to be the greatest uranium-related health risk to people on the Navajo reservation today. To continue mitigating this risk, EPA and NNEPA officials told us they plan to continue the work they conducted under the 2008 5-year plan, but they do not know how many homes they will ultimately need to assess and replace since there is no comprehensive source for this information. According to NNEPA officials, the agency has a backlog of more than 100 homes where residents have requested testing. A NNEPA official familiar with the work told us the agency expects to address this backlog in the 2014 5-year plan period. The official also told us that the number of requests continues to increase significantly as more people become aware of the agencies' efforts to assess houses and structures, in part, through outreach conducted by NNEPA. NNEPA is also responsible for communicating the results of completed assessments to residents when those results indicated homes were safe. A NNEPA official told us that, as of February 2014, the agency had communicated the results of fewer than half of the completed assessments, in some cases, because the agency was waiting for EPA to provide the results, and that the agency will continue to address this backlog moving forward. Overall, EPA officials said that they expect the total number of homes needing replacement will decrease in the 2014 5-year plan period since they believe the homes most likely to be contaminated have been addressed through earlier efforts.

Time frame: According to EPA officials involved, there is no end date for this work. They expect that the work assessing and potentially cleaning up contaminated houses to continue into the future as long as NNEPA continues to receive requests.

Cost: Given the unknown number of homes needing assessment and cleanup, EPA has not developed an overall cost estimate. However, EPA officials told us they typically spend $6,000 on a detailed assessment and from $80,000 to $300,000 to demolish a home and either provide financial compensation or a replacement home.

Assess and Treat Health Conditions and Conduct Health Research

Federal agencies do not have concrete plans or available funds to begin additional health studies on the Navajo reservation, but the Indian Health Service (IHS) will continue its work and the Agency for Toxic Substances and Disease Registry (ATSDR) and its partners will continue the Navajo Birth Cohort Study, spending up to $10 million from 2013 to 2018.

Scope of work: IHS and ATSDR officials told us they intend to include short-term health treatment and assessment efforts in the 2014 5-year plan, and they are developing overall goals for health research related to exposure to uranium on the Navajo reservation as well. In the short-term, beyond health care delivery to patients, IHS plans to continue its efforts under the uranium health-screening program that it established under the 2008 5-year plan. With 3.5 staff members, however, this is a small-scale effort. In addition, the Navajo Birth Cohort Study is expected to continue. Regarding other long-term research studies, ATSDR officials told us they do not have specific plans or funding to initiate additional uranium-related health research studies at this time. Members of the Navajo Nation, researchers, and other stakeholders have repeatedly called for a long- term epidemiological study of the effects of nonoccupational exposures to uranium in the communities that have lived closest to former mines and processing sites in order to gain a better understanding of how these communities have been and continue to be affected.[92] IHS officials told us they have been limited in their ability to identify or plan potential studies because conducting research studies is not a part of IHS's mission, and dedicated funding is not available for such efforts. ATSDR officials told us that their agency infrequently funds such studies; for example, it is currently conducting two long-term epidemiological studies across the country in addition to the Navajo study.

Time frame: IHS officials said they expect to extend the uranium health screening program into the future, at least through the 2014 5-year plan period. In 2013, ATSDR extended its agreement with the University of New Mexico and the Navajo Nation for the Navajo Birth Cohort Study by 5 years, until 2018.

Cost: ATSDR has committed to spending up to $10 million on the Navajo Birth Cohort Study from 2013 to 2018, but overall costs for potential future studies are unknown.

End Notes

[1] Uranium is a naturally occurring radioactive element found at elevated levels across the Navajo reservation and the entire Colorado Plateau. We are using the term "contamination" to refer to the presence of uranium in soil and water at the point where it may present a health risk to humans.

[2] Because the uranium mines in the region were located on or near Navajo lands, this report is focused primarily on the Navajo Nation. We discuss effects on the Hopi Tribe from a former landfill known as the Tuba City Dump, located largely on Hopi land.

[3] Pub. L. No. 95-604, 92 Stat. 3021 (1978).

[4] EPA, et al., *Health and Environmental Impacts of Uranium Contamination in the Navajo Nation: Five-Year Plan as requested by House Committee on Oversight and Government Reform*, June 9, 2008.

[5] Pub. L. No. 96-510 (1980), *codified as amended at* 42 U.S.C. §§ 9601-9675.

[6] Federal and tribal agencies estimate the number of homes and people without regulated drinking water on the Navajo reservation through different methodologies. As a result, their estimates of these numbers vary.

[7] In August 2010, EPA and BIA entered into an Administrative Settlement Agreement and Order on Consent for BIA to conduct a remedial investigation and feasibility study at the Tuba City Dump. BIA's work under the settlement agreement is subject to EPA's approval, and EPA will make the decision regarding a cleanup remedy.

[8] ATSDR was established within the Public Health Service of the Department of Health and Human Services. ATSDR is supported by the Centers for Disease Control and Prevention (CDC) and located within CDC's Office of Noncommunicable Disease, Injury and Environmental Health.

[9] EPA, et al., *Federal Actions to Address Impacts of Uranium Contamination in the Navajo Nation: Five-Year Plan Summary Report*, January 2013.

[10] GAO, *GAO Cost Estimating and Assessment Guide: Best Practices for Developing and Managing Capital Program Costs*, GAO-09-3SP (Washington, D.C.: Mar. 2009); and *GAO Schedule Assessment Guide: Best Practices for Project Schedules*, GAO-12-120G (Washington, D.C.: May 2012).

[11] Early mining on the reservation was for the mineral vanadium, which was used in steel production, and uranium was extracted from the vanadium mill tailings.

[12] The more recent technique for extracting uranium in the United States is called in situ recovery, which aims to extract uranium with less surface disturbance. According to NRC, this technique involves extracting uranium from aquifers that naturally contain uranium and other hazardous constituents by injecting oxygenated water and carbon dioxide or sodium bicarbonate hundreds of feet underground to dissolve uranium located in a subsurface ore body. Once dissolved, the water and uranium mixture is pumped to the surface. The primary risk associated with in situ recovery is the potential for contamination of nearby groundwater outside of the ore zone. When operations cease, companies remove and stabilize hazardous constituents found in the groundwater that may have been disturbed from the ore body by the operations. How long it will take to restore an in situ recovery site after production ceases is variable, and restoration to background levels may not be achievable for every constituent, according to NRC officials with whom we spoke.

[13] House Committee on Oversight and Government Reform, *The Health and Environmental Impacts of Uranium Contamination In the Navajo Nation*, 110th Cong., 1st sess., 2007.

[14] Background radiation is the radiation that is emitted from naturally occurring materials, such as from radium that is present in nearly all soils and rocks, or from other elements found in the earth's crust, including thorium and potassium. Background radiation varies all over the world, depending on the makeup of soils and rocks in the local area and on the elevation of the local area, which affects the amount of exposure to cosmic radiation. EPA considers radiation levels of more than 2-times background levels as evidence of an observed hazardous release that may require further investigation under CERCLA.

[15] Pub. L. No. 96-510 (1980), *codified as amended at* 42 U.S.C. §§ 9601-9675.

[16] Pub. L. No. 95-604, 92 Stat. 3021 (1978).

[17] Courts have held responsible party liability under CERCLA to be strict, joint and several, and retroactive. Under strict liability, a party may be liable for cleanup even though its actions were not considered negligent. Because liability is joint and several, when the harm done is indivisible, one party can be held responsible for the full cost of the remedy even though other parties may have contributed to the release of hazardous

substances at the site. Retroactive liability means that liability applies to actions that took place before CERCLA was enacted.

[18] Specifically, a federal district court has ruled that the federal government, as holder of title to the land on the Spokane Indian Reservation and entity responsible for approving and enforcing the uranium mine lease, is liable under CERCLA for the restoration of land damaged by uranium mining. *See United States v. Newmont USA Ltd*, 504 F.Supp.2d 1050, 1075 (E.D. Wash. 2007) ("[T]he drafters of CERCLA intended that land held by the United States in trust for Indians be treated the same as land owned in fee simple by the United States").

[19] Under CERCLA, federal agencies are generally not authorized to respond to a release of "a naturally occurring substance in its unaltered form, or altered solely through naturally occurring processes or phenomena, from a location where it is naturally found." 42 U.S.C. § 9604(a)(3). The burden is on the responsible party to prove that the government lacks the authority to respond to any particular release for this reason. E.g., *United States v. W.R. Grace - Conn.*, 280 F.Supp.2d 1149, 1175 (D. Mont. 2003). Releases resulting from mining activity that relocates the naturally occurring substances are not subject to the statutory limitation. E.g., *Monarch Greenback v. Monticello Insurance*, 118 F.Supp.2d 1068, 1080 (D. Idaho 1999). In this report, we use the term "contamination" to refer to releases of hazardous substances other than those covered by the statutory exemption. We make no findings herein as to whether any particular release is subject to the limitation.

[20] A site evaluation includes a site investigation and an engineering evaluation/cost analysis.

[21] CERCLA established the Superfund Trust Fund to support Superfund program activities. Historically, the Superfund Trust Fund received revenue from four major sources: taxes on crude oil and certain chemicals, as well as an environmental tax assessed on corporations based on their taxable income; transfers via appropriations from the general fund of the Treasury; fines, penalties, and recoveries from potentially responsible parties; and interest earned on the balance of the Superfund Trust Fund. As of 2011, the Superfund Trust Fund's primary source of revenue is the transfer from the general fund of the Treasury. EPA's Superfund program receives annual appropriations from the Superfund Trust Fund.

[22] 40 C.F.R. § 300.425(b)(1). In placing sites on the NPL, EPA uses a hazard ranking system to determine a site's relative threat to human health and the environment based on potential pathways of contamination and the availability of alternative state or federal programs that could clean up the site, among other things. EPA considers risks to human health and the environment, the urgency of the need for response, and program management factors, such as projected costs to the Superfund program and the timing of funding needs.

[23] There is a fifth former uranium mill site that is not located on the Navajo reservation but is located on private land, surrounded by Navajo land, directly across from the Northeast Church Rock mine. In contrast to the four mill sites on the reservation that are UMTRCA Title I sites, and which DOE was required to remediate, this mill site is an UMTRCA Title II site, which the private licensee, not DOE, is required to remediate. NRC licenses the site, subjecting it to NRC's decontamination, decommissioning, and reclamation standards, and the site is listed on the NPL. EPA issued an administrative order to the site's former operator to undertake a CERCLA remedial action. When this mill site is remediated to NRC standards the site will be transferred to DOE, which will perform long-term surveillance and maintenance.

[24] The Rare Metals Corporation and its successor, the El Paso Natural Gas Company, operated the Tuba City uranium processing site from 1956 to 1966. DOE refers to this former processing site as the Tuba City site, but in order to distinguish it from the other contaminated sites in the Tuba City area discussed in this report, we refer to the Tuba City former uranium processing site as the Rare Metals site.

[25] DOE and the Navajo Nation entered into a cooperative agreement in 1983 to facilitate the surface remedial actions at the sites on the reservation. This agreement expired in 1998, and a new cooperative agreement was initiated for groundwater cleanup that remains in effect.

[26] According to DOE officials, no uranium mines met the statutory definition of a vicinity property.

[27] The Navajo Abandoned Mine Lands Reclamation Program conducts reclamation work on uranium mines using Abandoned Mine Reclamation Fund money. Under the Surface Mining Control Reclamation Act, the Navajo Nation applies for grants from the Abandoned Mine Reclamation Fund to reclaim non-coal lands and waters and develop infrastructure in communities affected by mining activities. *See* 30 U.S.C. §§ 1235(k), 1240a(b); 53 Fed. Reg. 17186 (May 16, 1988); 59 Fed. Reg. 49178 (Sept. 27, 1994).

[28] We have reported that federal agencies do not have a consistent definition of an abandoned hardrock mine, including uranium mines. For example, some agencies count each mine feature as a separate mine where others group all features associated with one mine claim as a mine. EPA and NNEPA organize the abandoned uranium mines on the Navajo reservation by discrete mine claim. In this report, we are using this definition of an abandoned mine and therefore are reporting on the number of claims. For readability, we refer to these mine claims as mines. See GAO, *Uranium Mining: Opportunities Exist to Improve Oversight of Financial Assurances*, GAO-12-544 (Washington, D.C.: May 17, 2012); and *Hardrock Mining: Information on Abandoned Mines and Value and Coverage of Financial Assurances on BLM Land*, GAO-08-574T (Washington, D.C.: Mar. 12, 2008).

[29] The UMTRCA cooperative agreement is authorized by 42 U.S.C. § 7915(a).

[30] In addition to establishing targets under eight objectives, the 2008 5-year plan included a more detailed discussion of each objective which, in some cases, identified specific actions the federal agencies planned to take and

interim milestones the agencies planned to meet. This table shows the targets established in the 5-year plan but does not identify all actions contained in the plan.

[31] EPA undertook other interim removal actions, such as constructing fences, at two additional sites.

[32] EPA officials said this prioritization is based on the best information available and that they plan to add more mines to the list as necessary as they gain additional information about how the land surrounding the mines may be used by local residents.

[33] In 2013, ATSDR extended its agreement with the University of New Mexico and the Navajo Nation for the Birth Cohort Study by 5 years, until 2018.

[34] 5 C.F.R. § 1320.10(b). Under the Paperwork Reduction Act, the Director of OMB must review and approve agencies' proposed collections of information. OMB had 60 days from March 30, 2012—the date ATSDR submitted the proposed information collection for the study—to approve, disapprove, or request changes to the information collection. However, OMB did not request supplemental information from ATSDR until July 30, 2012, and did not approve the collection until February 14, 2013. Because OMB missed the 60-day regulatory deadline, ATSDR was allowed to request an OMB control number to validly collect the information for a year. 5 C.F.R. § 1320.10(c). However, the agency did not make such a request since the study would take more than a year to collect the information. OMB staff identified a number of factors influencing the length of the review process, including the complexity of the information included in the review and the large number of study partners participating in discussions related to the review. They also noted that ATSDR may have reduced the review time by submitting its documents for approval earlier in the process.

[35] Pub. L. No. 111-5, 123 Stat. 115 (2009).

[36] Pub. L. No. 111-8, Div. C, Tit. III, 123 Stat. 524, 617-18 (2009). DOE used the existing UMTRCA cooperative agreement with the Navajo Nation to provide most of the funding to NNEPA, which led the cleanup efforts.

[37] According to NNEPA officials, they scanned 14 structures for radiation and identified 2 that were contaminated and needed remediation. Remediation of the 2 structures was completed in March 2014, and, as of April 2014, NNEPA had identified one additional structure that required scanning.

[38] Committee reports accompanying the fiscal year 2012 and 2013 appropriation acts directed that $2 million of funds appropriated to ATSDR be used to continue epidemiological studies of the health conditions caused by exposures to uranium released from mining and milling operations on the reservation.

[39] In 2011, EPA received $13.3 million from the bankruptcy of Tronox, Inc., a potentially responsible party as the successor to the Kerr-McGee Corporation, which operated contaminated sites on the reservation. Of the settlement funds, EPA received $12 million to conduct or finance response actions at 49 sites on or near the Navajo reservation, including mines. EPA received $1.3 million to conduct or finance response actions at Quivira mine. EPA reported that they did not spend the Tronox bankruptcy funds on the interim time-critical removal action at the Quivira mine since the other former operator conducted and financed the action under an administrative order.

[40] The federal agencies' January 2013 summary report of work accomplished under the 2008 5-year plan reported that the agencies had spent $110 million during the 5-year plan period; we are reporting a different amount because agencies provided us with updated data, and we are reporting it in 2013 constant dollars. According to agency officials, these figures represent the agencies' direct costs and do not include funds for salaries or benefits.

[41] NRC officials said that, given the high level of interest in the site, they believe it is likely that a petition for a hearing on the proposed action for the license amendment request will be filed, and that the hearing process may take several years.

[42] According to EPA's assessment of removal options, this is an order of magnitude estimate rather than a specific, point estimate. Due to uncertainties, total cleanup costs could fall within a range from minus 30 percent to plus 50 percent of the $44 million.

[43] BIA is also responsible for implementing the approved work plan by awarding a contract that incorporates the requirements addressed in the work plan, and ensuring that its contractors properly perform the contract. Under the settlement agreement, BIA is responsible for ensuring its contractors comply with the agreement.

[44] On June 1, 2011, BIA issued a task order for the RI/FS in the amount of $1,994,577, under an indefinite delivery indefinite quantity contract.

[45] DOE has funded work that has contributed to the understanding of conditions at the site as well. Specifically, DOE provided funding to NNEPA to drill seven groundwater characterization wells between the Rare Metals site, located about 4 miles away, and the Tuba City Dump. According to DOE, sampling from the wells demonstrated that no groundwater contaminants were migrating from the Rare Metals site toward the dump.

[46] Both tribes have sued the federal government over the Tuba City Dump and Highway 160 site. Specifically, in 2010, the Navajo Nation intervened in an appeal a potentially responsible party for the Rare Metals site filed challenging a federal district court's dismissal of its lawsuit over the sites. The United States Court of Appeals for the District of Columbia affirmed the dismissal. *See El Paso Natural Gas Co. v. United States*, 632 F.3d 1272 (D.C. Cir. 2011). The tribe then intervened in a subsequent lawsuit the potentially responsible party brought, which the district court also dismissed. *El Paso Natural Gas Comp. v. United States*, 774 F.Supp.2d 40 (D.D.C. 2011). The tribe and potentially responsible party subsequently brought additional legal claims

against the government, which the district court also dismissed. *El Paso Natural Gas Co. v. United States*, 847 F.Supp.2d 111 (D.D.C. 2012). The tribe and potentially responsible party appealed this decision, and the United States Court of Appeals for the District of Columbia affirmed the dismissal of the claims concerning the dump but allowed one of the claims concerning groundwater remediation at the Highway 160 site to be heard by the district court. *El Paso Natural Gas Company v. United States*, No. 12-5156, 12-5157, _ F.3d _ (D.C. Cir. 2014). The Hopi Tribe filed a complaint against BIA in Hopi Tribal Court. *See Hopi Tribe v. Bureau of Indian Affairs*, No. 2011-cv-0107 (Hopi Tribal Court).

[47] Under the terms and conditions of the contract, BIA issued a cure notice to its contractor on November 9, 2012. The cure notice stated that BIA would terminate the contract within 10 days if the contractor did not cure the conditions stated in the notice that were endangering performance of the contract.

[48] GAO, *Acquisition Planning: Opportunities to Build Strong Foundations for Better Services Contracts*, GAO-11-672 (Washington, D.C.: Aug. 9, 2011).

[49] The settlement agreement allows BIA to request an extension to deadlines in the work plan schedule from EPA in advance of a deadline and requires EPA to approve a timely extension request when good cause exists for the extension.

[50] Under the settlement agreement, BIA is liable for specified stipulated penalties per violation per day for failure to comply with any of the agreement's requirements, including failure to perform work as required in the approved work plan and failure to submit timely or adequate reports or documents specified by EPA. EPA reviewed the reasons BIA gave for missing deadlines and considered BIA's past work, as well as its proposal for future work, to determine which missed deadlines were justifiable. According to EPA documents, at least 966 days of missed deadlines in 2012 were unjustifiable, including 54 days due to BIA coordination and contractor performance. EPA held the stipulated penalties in abeyance for the missed deadlines it determined to be unjustifiable.

[51] According to EPA, the issuance of penalties was held in abeyance because of (1) BIA's efforts to correct deficiencies in its management that had led to delays, reattain schedule progress, and comply with an optimized schedule to make up for some delays and (2) project delays that can result from the process of issuing penalties. EPA has reserved the right to issue penalties for the deadlines BIA missed in 2012. Based on EPA documents, the penalties appear at a minimum to be between $443,000 and $887,500.

[52] According to EPA documents, the 966 days of missed deadlines translated into a few weeks' delay because the deadlines were for interim products or actions, and there were multiple concurrent violations.

[53] The work plan deadline for submittal of the draft Remedial Investigation report was on August 23, 2013. BIA requested an extension of this deadline on September 3, 2013, and submitted the draft report on the new deadline of November 15, 2013.

[54] A schedule is comprehensive if it includes all activities for both the government and its contractors necessary to accomplish a project's objectives and realistically reflects how long each activity will take; is well-constructed if all its activities are logically sequenced; is credible if it reflects the order of events necessary to achieve a final outcome and data about risks are used to predict a level of confidence in meeting the project's completion date; and is controlled if it is updated by trained schedulers using actual progress to realistically forecast dates. See GAOand GAO-12-120G.

[55] BIA submitted an extension request in September 2013 and another in October 2013. EPA did not initially approve the September 2013 request because it found that BIA had not demonstrated good cause for the extension as required by the settlement agreement. However, EPA did not respond to the September request within the time period required by the settlement agreement so EPA was deemed to have concurred with the extension. EPA did not find the October 2013 request to be justified but did find the October 2013 government shutdown to constitute good cause to adjust the schedule.

[56] BIA and its contractor have estimated the costs for two remedial actions: (1) a low-cost option that involves closing the landfill in place and installing a cap and cover system and (2) a high-cost option that entails excavating the waste and contaminated material and disposing of it offsite.

[57] GAO

[58] EPA has already selected cleanup actions at 2 of the 43 highest priority mines: Skyline mine, where the cleanup action was selected and has been conducted, and Northeast Church Rock mine, where EPA has selected the removal action but it has not been completed. According to EPA officials, since they conducted the cleanup at Skyline mine to be protective of human health and the environment in the long term, they consider the cleanup action to be complete. The Navajo Nation, however, considers the cleanup to be temporary given its current position regarding removing all mine waste from the reservation.

[59] GAO

[60] GAO

[61] Specifically, this report is required to identify the locations of abandoned uranium mines across the country, including on the Navajo reservation; describe the risks associated with the mines; prioritize the remediation and reclamation of the mines; analyze the cost of remediating and reclaiming each mine; and describe the status of any efforts to remediate and reclaim the mines. Pub. L. No. 112-239, § 3151 (2013). The statutory

reporting deadline is July 2, 2014. According to DOE, the agency met with EPA and Navajo agencies, as required, to ensure that all mines on Navajo lands are incorporated in the report.

[62] In 2012, IHS estimated that there were approximately 5,500 homes on the entire Navajo reservation without piped water.

[63] Specifically, according to DOE officials, lower volumes of groundwater present underneath the Rare Metals site have resulted in reduced efficacy of the pump-and-treat system. DOE has operated the system since 2002, and it has not observed significant or widespread decreases in the concentration of contaminants under the site. DOE officials estimate it could take 400 years to achieve water quality standards with the current system. Similarly, at Shiprock, DOE has run a pumping system for the past 10 years, but it has not decreased overall contamination levels, although the extent of one plume has decreased, and many previously existing contaminant seeps have gone dry. DOE estimated in 2002 that it would need to operate portions of the pumping system for up to 60 years; however, DOE officials said that they have not recalculated that number in light of actual performance of the system.

[64] The more thorough assessments, as we refer to them here, include (1) a site evaluation, including an engineering evaluation/cost analysis, to support a non-time critical removal action or (2) an RI/FS to support a remedial action.

[65] We created this rough estimate based on the following assumptions: (1) EPA would need to fully fund removal actions at 21 of the highest priority mines that most likely will not have a potentially responsible party available to contribute to the costs; (2) the costs for these removal actions would average, at a minimum, what EPA spent on the removal action conducted at Skyline mine, or about $7 million; and (3) EPA Region 9's Superfund removal budget funding levels from the 2008 5-year plan period continue into the future.

[66] In addition, if the federal government is a potentially responsible party at any of these highest priority sites, it is also liable for cleanup costs, thereby increasing the amount of federal funds necessary for cleanup.

[67] Further, a NNEPA water program official explained in a presentation that, even if the residents choose to haul water from a regulated source, CDC found that the cisterns used by many residents to store their water at their homes contained bacteria, which could contaminate clean water hauled and then stored in the cistern.

[68] This estimate is based on the same assumptions as our time frame estimate.

[69] According to EPA officials, Skyline mine was a small-sized mine, and these officials said they believe it represents the low end of costs for these mines. In addition, EPA did not incur transportation costs at Skyline mine because the agency disposed of the waste in a newly constructed cell located on-site.

[70] A senior Navajo Area IHS official told us that the agency funded 7 of the 36 projects in fiscal year 2013, after the 2008 5-year plan was complete. Therefore, the remaining 28 economically feasible projects are estimated to cost about $29 million.

[71] DOE created its estimate to contribute to the agency's environmental liability estimate and includes anticipated work associated with the four former processing sites on the Navajo reservation, DOE's cooperative agreements with the Navajo Nation and Hopi Tribe, and the expected disposal of waste from Northeast Church Rock mine at a former processing site just off the reservation.

[72] The Safe Drinking Water Act Amendments of 1996 established the Drinking Water State Revolving Fund, which provides annual funding to states to finance projects for publicly and privately owned drinking water treatment systems. Specifically, states use the funding to provide low- or no-interest loans to communities or utilities, which, when repaid, allows the funding of future loans for additional projects. Tribes receive funding through a set- aside account that represents a percentage of the total appropriation for the fund.

[73] According to EPA officials, EPA Region 9 has used the Superfund Trust Fund to pay for removal actions through its removal program.

[74] GAO, *Standards for Internal Control in the Federal Government*, GAO/AIMD-00-21.3.1 (Washington, D.C.: November 1999).

[75] 79 Fed. Reg. 20910 (Apr. 14, 2014); *In re: Tronox, Incorp., et al*, Case No. 09-10165 (Bankr. S.D. N.Y.).

[76] EPA Region 9's Office of Air and Radiation developed the list of potentially contaminated structures, and officials from the region's Superfund program told us they were not aware of the list until the Navajo Nation presented it at a 2007 congressional hearing on Navajo uranium contamination. A NNEPA program official told us that, subsequently, in assessing homes on the list for contamination, one homeowner asked her if she had come to retrieve an old radon testing canister. The home owner presented the official with the canister—stamped with EPA's name and logo—and told her that EPA had placed the canister in her home in the 1970s or early 1980s but had never returned to communicate the results.

[77] In recent years, some mining companies have expressed a renewed interest in uranium mining on or near the Navajo reservation. For example, since 2005, one company has pursued NRC approval to renew its license for uranium mining on four sites in New Mexico that are near the reservation, according to NRC officials we spoke with. In 2012, the company also signed an agreement with the Navajo Nation allowing it to cross tribal trust land in order to access a site where it has been conducting exploration; however, the agreement specifically states that the company cannot begin mining uranium on that site until legacy waste at that site and an adjacent site has been cleaned up.

[78] 42 U.S.C. § 7914(f)(2);10 C.F.R. § 40.27(b), (c).

[79] In October 2011, we reported on EPA's efforts for promoting environmental justice, that is, the fair treatment and meaningful involvement of all people in developing, implementing, and enforcing environmental laws, regulations, and policies. The concept of environmental justice is based on the belief that communities with large numbers of minority or low-income residents frequently shoulder a disproportionate share of environmental and health risks. Many of these communities are located in areas within close proximity to sources of pollutants that can adversely affect both the environment and human health. See GAO, *Environmental Justice: EPA Needs to Take Additional Actions to Help Ensure Effective Implementation,* GAO-12-77 (Washington, D.C.: Oct. 6, 2011).

[80] The project involves relocating 16 million tons of tailings and tailings contaminated soil. DOE estimates that the project will be completed in 2025, at a total project cost of about $1 billion.

[81] The U.S. government has issued formal apologies for past events. For example, in Pub. L. No. 103-150, 107 Stat. 1510 (1993), Congress apologized to Native Hawaiians on behalf of the people of the United States for the overthrow of the Kingdom of Hawaii with the participation of agents and citizens of the United States. In addition, in the Radiation Exposure Compensation Act, Congress apologized on behalf of the nation to underground uranium miners and their families, among others, for the hardships they endured. Pub. L. No. 101-426, 104 Stat. 920 (1990).

[82] GAO, *Results-Oriented Government: Practices That Can Help Enhance and Sustain Collaboration among Federal Agencies*, GAO-06-15 (Washington, D.C.: Oct. 21, 2005).

[83] Other key practices we identified in GAOthat can enhance and sustain interagency collaboration are: define and articulate a common outcome; identify and address needs by leveraging resources; establish compatible policies, procedures, and other means to operate across agency boundaries; develop mechanisms to monitor, evaluate, and report on the results; reinforce agency accountability for collaborative efforts; and reinforce individual accountability for collaborative efforts.

[84] In an April 2013 letter to the Executive Directors of NNEPA and the Navajo Nation Division of Natural Resources, which houses the Abandoned Mine Lands Reclamation program, the Navajo Nation Department of Justice encouraged the two agencies to continue to work cooperatively at sites contaminated by past uranium mining or processing while stating its position that NNEPA exercises regulatory authority over all actions taken by any other Navajo Nation agency at these sites.

[85] We have assessed environmental cleanup projects against the criteria found in our cost estimating and assessment guide and our schedule assessment guide and have reported on the results of these assessments. See, for example, GAO, *Nuclear Weapons: Factors Leading to Cost Increases with the Uranium Processing Facility*, GAO-13-686R (Washington, D.C.: July 12, 2013); *Nuclear Waste: Actions Needed to Address Persistent Concerns with Efforts to Close Underground Radioactive Waste Tanks at DOE's Savannah River Site*, GAO-10-816 (Washington, D.C.: Sep. 14, 2010); and *Department of Energy: Actions Needed to Develop High-Quality Cost Estimates for Construction and Environmental Cleanup Projects*, GAO-10-199 (Washington, D.C.: Jan. 14, 2010).

[86] GAO

[87] GAO

[88] GAO-12-120G.

[89] We determined the overall assessment rating by assigning each individual rating a number: Not Met = 1, Minimally Met = 2, Partially Met = 3, Substantially Met = 4, and Met = 5. Then, we took the average of the individual assessment ratings to determine the overall rating for each of the four characteristics. The resulting average becomes the Overall Assessment as follows: Not Met = 1.0 to 1.4, Minimally Met = 1.5 to 2.4, Partially Met = 2.5 to 3.4, Substantially Met = 3.5 to 4.4, and Met = 4.5 to 5.0.

[90] GAO

[91] We determined the overall assessment rating by assigning each individual rating a number: Not Met = 1, Minimally Met = 2, Partially Met = 3, Substantially Met = 4, and Met = 5. Then, we took the average of the individual assessment ratings to determine the overall rating for each of the four characteristics. The resulting average becomes the Overall Assessment as follows: Not Met = 1.0 to 1.4, Minimally Met = 1.5 to 2.4, Partially Met = 2.5 to 3.4, Substantially Met = 3.5 to 4.4, and Met = 4.5 to 5.0.

[92] In April 2013, Radiation Exposure Compensation Act Amendments of 2013 were introduced in the House and Senate. The bills, if enacted into law, would require the Secretary of Health and Human Services to establish and administer a grant program through the National Institutes of Health for the study of the epidemiological impact of uranium mining and milling among nonoccupationally exposed individuals, including family members of miners and millers. As of February 2014, the bills had not been enacted into law. H.R. 1645, 113th Cong. (2013); S. 773, 113th Cong. (2013).

In: Uranium Contamination in the Navajo Nation
Editor: Eugene Stepp
ISBN: 978-1-63321-678-5

Chapter 2

HEALTH AND ENVIRONMENTAL IMPACTS OF URANIUM CONTAMINATION IN THE NAVAJO NATION: FIVE-YEAR PLAN*

Bureau of Indian Affairs, Department of Energy, Nuclear Regulatory Commission, Environmental Protection Agency and Indian Health Service

EXECUTIVE SUMMARY

Mine operators extracted nearly four million tons of uranium ore from 1944 to 1986 under lease agreements with the Navajo Nation. As a result, uranium mining has left the Navajo Nation with a legacy of over 500 abandoned uranium mines (AUMs), four inactive uranium milling sites, a former dump site, contaminated groundwater, structures that may contain elevated levels of radiation, and environmental and public health concerns.

The House Committee on Oversight and Government Reform asked the Bureau of Indian Affairs (BIA), Department of Energy (DOE), Nuclear Regulatory Commission (NRC), Environmental Protection Agency (EPA), and Indian Health Service (IHS) to develop a coordinated five-year plan with specific milestones to address the following objectives:

- Assessment of structures and water sources that are likely to be contaminated;
- Cleanup of structures found to be contaminated above safe levels;
- Provision of alternate water supplies for residents consuming contaminated water;
- Tiered assessment of abandoned mines, with more detailed assessments of those most likely to pose environmental or health problems;
- Cleanup of the Northeast Church Rock mine site and additional high-priority abandoned mine sites;
- Cleanup of the Tuba City Highway 160 site;

* This is an edited, reformatted and augmented version of a plan requested by the House Committee on Oversight and Government Reform, issued June 9, 2008.

- Cleanup of the Tuba City Dump;
- Remediation of groundwater contamination at three former mill sites; and
- Conduct of one or more case control studies of health risks faced by individuals residing near mill sites or abandoned mine sites.

The agencies have worked together to address public health and environmental impacts from historical uranium mining on the Navajo Reservation and continue to consult with our Navajo Nation counterparts on the plan. The Navajo Nation Department of Justice provided guidance on federal consultation procedures on November 16, 2007, and February 12, 2008. In addition, four of the five agencies presented the draft Five-Year Plan and consulted with the Navajo Nation Resources Committee on April 18, 2008.

The agencies look forward to continued coordination and consultation with the Navajo Nation as we implement this plan. In addition, portions of this plan have the potential to affect lands on or adjacent to the Hopi Reservation. Because of the need for continued consultation with the Navajo Nation and Hopi tribe, this document and the action plans contained within are not final, and they will change as new information becomes available to the federal agencies preparing this plan.

The Committee also asked the agencies to identify cost estimates necessary to achieve our objectives. In developing this action plan, the agencies will follow existing agency programs and policies. Costs estimated for 2008 and 2009 are provided in the greatest detail since agency budgets for this period are in place and the costs are known with greater certainty than those for 2010-2012. Some activities and funding levels discussed in this report may not be included in agencies' budgets. Agencies will need to ensure that these activities and funding levels are consistent with other priorities and resource demands.

In 2008, agencies will make substantial progress on the most urgent problems, including:

- assess and clean up potentially contaminated structures;
- test water sources with suspected uranium contamination and provide assistance to affected residents;
- require mitigation of contaminated soils at the highest-risk mine -- Northeast
- Church Rock; and
- assess risk to drinking water supplies at the Tuba City Dump and implement any needed interim remedial actions.

By 2012, EPA expects to complete a tiered assessment of abandoned uranium mines and to complete soil cleanup of the Northeast Church Rock mine. BIA will lead the implementation of a final remedy for the Tuba City Dump, which is targeted for 2011. Throughout the five-year period, DOE Legacy Management will continue maintenance of contaminated groundwater treatment activities at three inactive uranium milling sites that have been remediated by DOE Environmental Management. The NRC, working closely with the Navajo Nation, will conduct a comprehensive safety and environmental review of any new application for a uranium recovery site. The Indian Health Service (IHS) will diagnose and treat known health conditions in eligible Indians and will work with others to assess the potential for case control studies.

The agencies have identified the following specific objectives for the five-year period from 2008 to 2012:

- *Assess and Remediate Contaminated Structures*: In coordination with the Navajo Nation, EPA will assess up to 500 structures on their land that are likely to be contaminated with radiation at elevated levels. Where contaminated structures are found that pose a health risk to current or future occupants, EPA will take appropriate response actions using its Superfund authority. Actions may include demolition of a structure or a portion of a structure, construction of a replacement structure, or other actions to protect current and future residents from exposure to elevated levels of radiation.
  The estimated cost for assessing 110 structures in 2008 is $650,000, and the estimated cost for remediating each contaminated structure is $100,000-$300,000 per structure.
- *Assess Potentially Contaminated Water Sources and Assist Affected Residents*: In spring 2008, EPA will test up to 70 rural Navajo water sources such as livestock wells, windmills, artesian wells and springs to determine if they contain safe levels of radiation or radionuclides. EPA will work with the Navajo Nation, the Indian Health Service, and other agencies to provide the results to the affected communities and develop plans to ensure access to clean and safe water for drinking. EPA will use Superfund authority to take appropriate action at water sources contaminated by mine waste and found to be endangering human health.
  EPA estimates the cost of assessing water sources and providing outreach through a Navajo EPA circuit rider in 2008 to be $250,000 and $150,000 per year in 2009. The Indian Health Service has estimated the cost of providing alternative water supplies for eligible Indians consuming water from 41 unregulated and potentially contaminated water sources to be up to $65 million, including planning, design, and construction costs. This amount is not accounted for in the IHS budget, and this estimate will need to be recalculated after more reliable techniques are used to determine contamination levels.
  These projects will impact approximately 730 permanently occupied Navajo homes that currently do not have piped water. The Navajo Nation may, but is not obligated to use grant funds provided by the Office of Surface Mining for construction of public facilities, including replacement water supplies, in mine- impacted areas. Office of Surface Mining expects to provide an estimated $34 million in grant funds to the Navajo Nation during the five-year period of this plan.
- *Assess and Require Cleanup of Abandoned Uranium Mines*: EPA and the Navajo Nation will work together to conduct a tiered assessment of abandoned mines, with more detailed assessments of those most likely to pose environmental or health problems. In 2007, EPA completed a geographic information system inventory and risk-ranking of the 520 known mines. In 2008, EPA will use this list and consult with the Navajo Nation to identify the highest- priority 200-250 mines for further evaluation. EPA expects to conduct on-site screening evaluations of these mine sites, at the rate of up to 50 per year through 2012. EPA expects to conduct more detailed assessments of up to 35 of the highest priority mines by 2012, at the rate of approximately seven sites each year.

High-priority mines will be candidates for Superfund enforcement. EPA will continue to pursue an enforcement-first policy, and will conduct a general search for Potentially Responsible Parties at all of the mines. EPA plans to issue information request letters to several major viable corporations (responsible parties) who owned or operated multiple mines by 2008, and EPA will continue to assist Navajo Department of Justice with Navajo-lead enforcement actions.

EPA estimates the cost of conducting a tiered assessment of mines to be $200,000 in 2008 and $150,000 in 2009.

- *Cleanup the Highest-Priority Mine Site*: The Northeast Church Rock Mine located near Gallup, New Mexico, is the highest priority cleanup on EPA's AUM ranking list. EPA will determine the soil remedy in 2008 and plans to require United Nuclear Corporation, now a subsidiary of GE, to perform a comprehensive Superfund removal action for cleanup of soils on the balance of the site. Cleanup is expected to take up to three years. Costs are expected to be provided by the responsible party.
- *Remediate Groundwater at the Inactive Uranium Milling Sites*: DOE's Office of Legacy Management will continue maintenance of existing groundwater treatment activities at the three inactive uranium ore processing sites and long- term surveillance and maintenance at four inactive milling sites, all of which have been remediated by DOE Environmental Management, throughout the 5-year period of this plan. Site surveillance and maintenance is estimated at $4,940,000 in 2008 and $3,178,000 in 2009.
- *Characterize the Highway 160 Site*: DOE will work with BIA, USGS, EPA, the Navajo Nation, and El Paso Natural Gas (EPNG), as appropriate, to understand the best approach for characterization and any required cleanup based on characterization of the site.

  There are currently three authorized courses of action related to additional characterization and remediation of the Highway 160 site: (1) EPA cleanup of the site as a Superfund removal action; (2) continued action by EPNG; (3) some combination of the two approaches.

  If EPA determines that an Imminent and Substantial Endangerment exists at this site, EPA can use Superfund enforcement authority to require the potentially responsible parties to conduct a response action. DOE will cooperate with that effort, as requested by the parties as appropriate. If El Paso Natural Gas works to effect a cleanup of this site, DOE will work with them to consider disposal of the contaminated materials at the DOE facility near Grand Junction.

  DOE has provided support to the preparation of a cost estimate for the remaining cleanup work based on the EPNG characterization report. The cost estimate would require additional characterization data that provides information on the chemical species and the depth of the contamination. Based on existing site knowledge, DOE estimates it would take several years to complete the work at the Highway 160 site.
- *Assess and Cleanup the Tuba City Dump*: The five federal agencies will work together with the Navajo and Hopi tribes in early 2008 to assess whether interim actions are needed prior to selection of a final remedy for the site. By August 2008, the Bureau of Indian Affairs plans to complete an assessment of the need and feasibility of conducting an interim measure to prevent contamination of nearby Hopi water supplies. If an imminent threat to water supplies is identified, the agencies will

determine the most appropriate authorities to achieve an interim remedy. These authorities might include a Superfund response or enforcement action.

As of 2008, committed funds include $916,000 to perform additional studies needed to determine the movement of contaminated groundwater, and potential threat to drinking water supply wells, irrigation water, springs, and seeps. These studies are also necessary in order to determine the need for, and scope of remediation.

- *Assess and Treat Health Conditions*: The Indian Health Service will continue to diagnose and treat known health conditions in eligible Indians, and also support a university-led Navajo Uranium Assessment and Kidney Health Project in 2008. IHS will also review existing databases to develop plans for improved cancer case surveillance, review water contamination data for potential future health studies, and develop plans to assess the prevalence of cancer and other health conditions for populations near AUM and inactive mill sites. Cost estimates for this objective are unavailable, but are projected to be achievable within current and projected agency appropriations.
- *Summary of NRC Actions*: The Nuclear Regulatory Commission (NRC) will consult with and continue conducting outreach to the Navajo Nation for new uranium recovery facility license applications that have potential cultural and resource impacts on Navajo lands. NRC will also continue to work with the Navajo Nation and other federal agencies on issues related to the cleanup of the Northeast Church Rock site and Tuba City contaminated properties. In addition, NRC will continue to consult with the Navajo Nation, as required by UMTRCA on DOE's remedial actions for cleanup of the four inactive uranium milling sites that will be selected and performed with concurrence of the Commission.

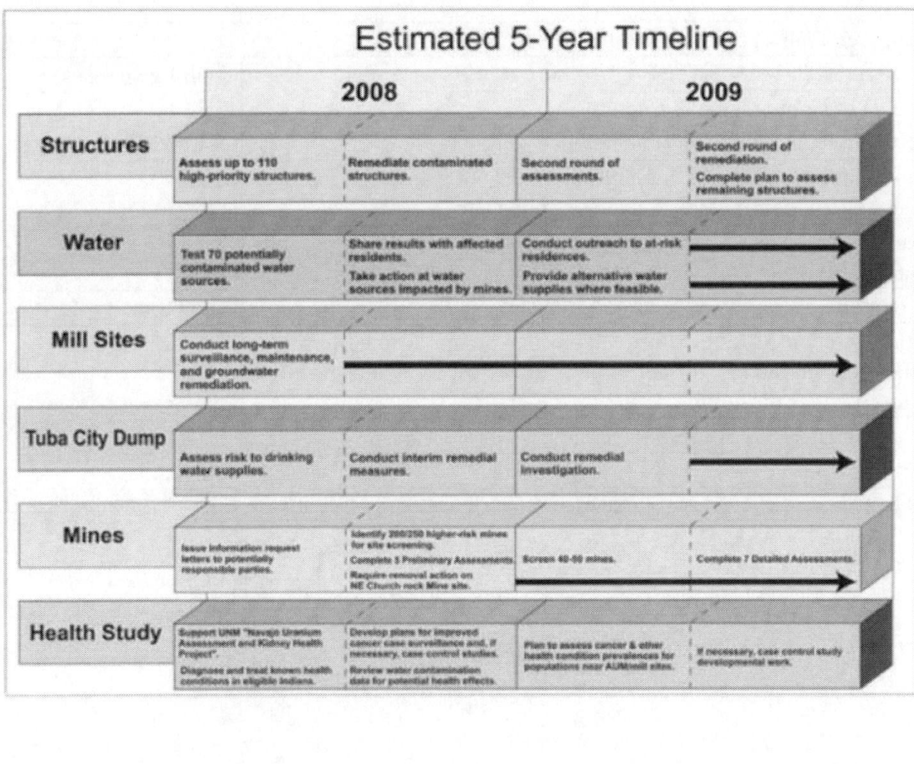

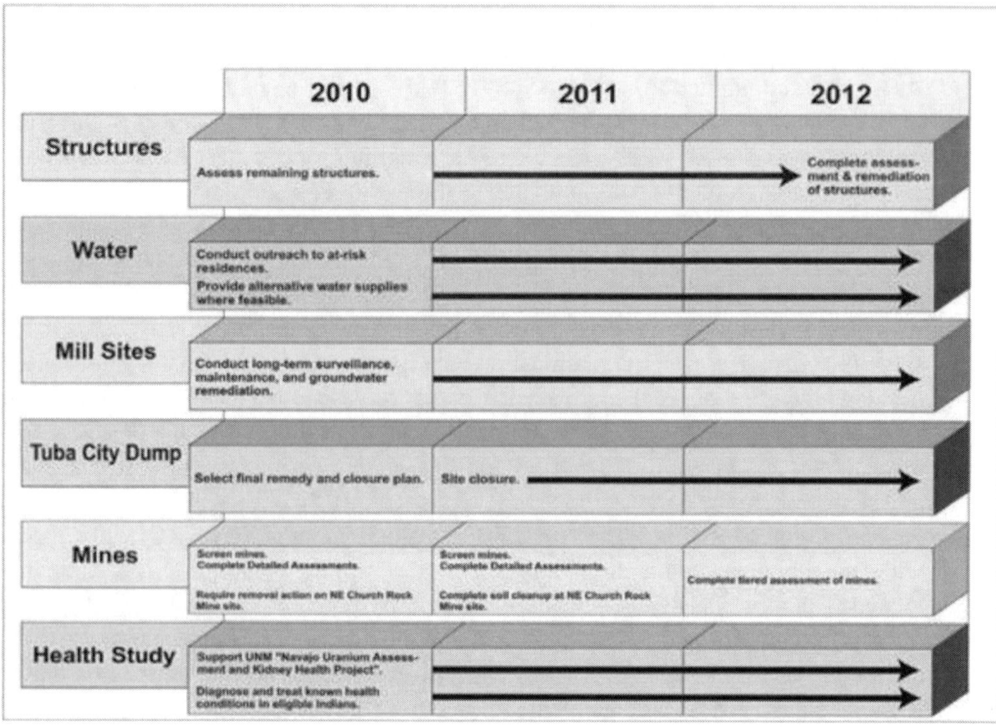

Agency Expenses Prior to 2008

Costs given in thousands of dollars	
Expenses 1997-2007	**Total Expenses**
Removal of Contaminated Structures	$ 84
Investigation and removals at abandoned mines	$ 13,000
Cleanup of the Northeast Church Rock mine site residential yards and additional high-priority abandoned mine sites	$ 2,290
Tuba City Dump assessment, closure, and monitoring activities and associated costs	$ 2,170
Remediation of four inactive uranium milling sites	$ 140,000
Medical screening examinations and special patient evaluations to former uranium miners, former millers, and "downwinders" from NV test bomb explosions.	$ 3,800
Contaminated Water Sources	$ 150
Total Expenses	**$ 161,494**

Note: The expenses above are estimated and may not capture all historic expenses incurred for these activities.

ASSESSMENT AND CLEANUP OF CONTAMINATED STRUCTURES

Agency: EPA

Background

Mine operators extracted nearly 4 million tons of uranium ore from 1944 to 1986 across the Navajo Nation. Uranium mining or milling waste has occasionally been used in the form of sediments for aggregate (foundations and stucco) and stones incorporated into the walls of structures, including homes. Structures may also be contaminated by the presence of radiological materials in outside dust and soil that was brought into the homes on shoes and clothing. These structures may be occupied as homes, periodically occupied for ceremonial purposes, or unoccupied. If the structures constructed with uranium mining or milling waste are occupied, there is a risk to the inhabitants from alpha radiation (e.g. radon gas), which is a potent lung carcinogen.

USEPA began assessment of structures for radiation on the Navajo Nation's lands in 1975, prior to the passage of Superfund legislation. At that time, EPA's Office of Radiation and Indoor Air (ORIA) and the predecessor to the Navajo Nation Environmental Protection Agency (NNEPA) investigated 37 structures located near the Monument Valley Mill Site and the Monument #2 Mine, both operated by Vanadium Corporation of America. Under its remediation authority under the Uranium Mill Tailings Radiation Control Act, DOE demolished four structures in Monument Valley that were found to have elevated radiation levels. As part of this action plan, EPA will reassess the remaining 33 structures identified in the 1975 survey to determine radiation levels.

In the 1990s, EPA and the Navajo Nation contacted local officials in 30 impacted communities and offered free radiation surveys. Based on feedback from the local officials, EPA surveyed 28 structures. Based on the results from these surveys, EPA removed two structures near Monument Valley and the Four Corners area at a cost of $84,000. EPA provided the affected families with compensation or a replacement structure. The other 26 structures did not have elevated radiation levels and are considered safe to occupy. These structures will not be reassessed as part of this action plan.

There may be additional structures that are not on the lists discussed above that may contain elevated levels of radiation and may present a risk if occupied. To assist the NNEPA in determining the priority for addressing these additional structures, EPA provided a list of over 500 structures located within ¼ mile of AUMs. This list was generated using EPA's GIS database, which was provided to NNEPA in August 2007. The status of these structures (whether they are homes, whether they are occupied, whether they are constructed with materials from mines, whether they have elevation radiation levels) is unknown. In addition, Navajo officials have indicated that there may be additional potentially contaminated structures that do not appear on the three lists discussed previously, or on the list of structures located within a ¼ mile of a mine.

Action Plan

In coordination with NNEPA, EPA plans to assess structures on the Navajo Nation's lands that are likely to be contaminated with radiation at elevated levels. Where contaminated structures are found that pose a health risk to current or future occupants, EPA will take appropriate response actions using our authority under the Superfund law. Actions may include demolition of a structure or a portion of a structure, construction of a replacement structure, or other actions to protect current and future residents from exposure to elevated levels of radiation.

Assessment and Remediation of Initial Lists of Structures (Phase I)

EPA has mobilized an assessment team to survey the 33 structures identified in NNEPA's desktop study and the 33 structures identified in the 1975 survey, for a total of 70 structures. The USEPA team is conducting radiological surveys of home and hogan interiors, and areas around these structures. The EPA team will also conduct radon sampling in the interior of structures. In addition to the 66 structures, EPA has committed to address other structures that may be of concern in the immediate vicinity of structures on the two lists. We estimate that this initial assessment will include approximately 110 structures in total. EPA anticipates that this assessment will be completed in Spring 2008; at an estimated cost of approximately $650,000 (costs include contractor costs, equipment costs, EPA personnel, and NNEPA personnel time).

In consultation with NNEPA, EPA will develop action levels and options for removal actions at structures with elevated radiation levels. EPA will begin Phase I removal actions in Summer 2008. While we cannot predict how many structures will require removal action, nor exactly what actions will be required, we estimate that remediation of an individual structure may cost approximately $100,000 to $300,000. In FY08, EPA can conduct approximately five removal actions at contaminated structures.

Residents will be offered temporary housing, along with pets and livestock if necessary, for the homesite remediation period. EPA will discuss the best options for each resident with them in person. Options include staying in a hotel or in homes of family members. EPA will provide reimbursement for accommodations as well as meals and incidental expenses (Federal per diem rates). EPA will reimburse, restore or replace any disposed, damaged or demolished portions of all structures and landscape type materials and will consult with the residents on all decisions. EPA will also conduct outreach to convey the risks of long-term exposure to any contaminants and encourage residents to meet with their primary care physician. EPA, ATSDR, and IHS are working together to provide professional education to the staff at multiple Navajo health clinics.

Assessment and Remediation of Additional Structures (Phase II)

In the course of working with NN Chapter officials and residents to identify and locate the Phase I structures, NNEPA staff identified additional structures not found on previous lists that may have been constructed with materials from mining operations and may have elevated radiation levels. NNEPA plans to interview residents and community health workers this summer to get more information on these structures. NNEPA has agreed to develop and prioritize a list of these additional structures and share it with EPA by Summer 2008. EPA and NNEPA will jointly develop a plan to assess the list of Phase II structures by the end of 2008. Options for assessment could include (1) EPA assessment of structures or (2) NNEPA assessment of structures, or (3) a combination of EPA and NNEPA assessment. Costs for these three options would depend on the number of structures on the list. A second round of assessments could begin in Spring 2009. Removal actions related to these assessments could begin in Fall 2009.

Plan for Remaining Structures of Potential Concern (Phase III)

Navajo Nation Chapter officials who have been notified of EPA's initial assessment efforts have stated that there may be more structures within their Chapters that may have been constructed with materials from mines. EPA recommends that NNEPA consult with these Chapters to identify any additional potentially-contaminated structures. In addition, USEPA has provided a list of 500 structures located within a ¼ mile of a mine, and recommends that NNEPA and Navajo chapter officials identify which of these might be contaminated. EPA would like NNEPA to create a prioritized list (Phase III) of structures identified by Chapters and structures of concern from the list of 500. EPA will work with NNEPA to develop a plan for assessment of these Phase III structures, with the goal of preparing this plan by the end of 2009. The plan would identify a prioritized list of Phase III structures to be assessed by NNEPA, EPA, or a combination of the two agencies.

EPA estimates that up to 500 structures may require assessment during Phases I, II, and III during the five-year period. Assessment costs per structure are approximately $6,000. It is possible that the Navajo Nation could elect to use a portion of their OSM 411 funding for these assessments.

Limitations

Tribal Consultation: EPA is working closely with NNEPA on community and cultural issues. USEPA has discussed with the NNEPA the process for assessing the Phase I structures identified in this action plan. USEPA will continue to work with NNEPA Superfund Program and residents to secure access to these structures and perform surveys. EPA will be consulting with NNEPA on potential actions including compensation and temporary relocation if demolition or partial removal is indicated, and future assessment activities.

Remediation, Transportation and Disposal Costs: While the volume of potentially radioactive material from structures is much less than that of the mines, transportation and disposal arrangements need to be discussed up front with the Navajo Nation as cost and capacity considerations. Specifically, EPA and NNEPA will need to identify where radioactive materials will be disposed. If materials are to be disposed of at a regulated radioactive waste disposal facility outside the Navajo Reservation, remediation costs will be significantly higher than the estimated $50,000-$100,000 per structure. Further, because the number of structures to be remediated is unknown, total remediation costs cannot be easily estimated.

Structures with high naturally occurring radon levels but low gamma radiation levels: If EPA identifies structures that are found to have high naturally occurring radon levels but low gamma radiation levels, then EPA will involve the Navajo Radon Program to provide abatement of these structures, which would be relatively low cost. These conditions would indicate radon that is not attributable to mining activity.

Land Use Issues: Based on existing policies, where contaminated structures or surrounding soils are found to pose a health risk to residents, EPA will take a removal action. If the entire area where a structure is located is found to pose a health risk, or a decision is made to build a replacement structure in another location, land use issues may arise that are beyond EPA's ability to address and will require the involvement of other Navajo agencies to resolve.

CONTAMINATED WATER SOURCES

Agency: EPA

Background

There are 197 public water systems on the Navajo Nation. These water systems serve 70% of the population on the Navajo Nation. In the past year 91% of these systems were in compliance with all health-based regulations, thereby exceeding the national compliance rate of 89.1%. The non-compliance was caused by the presence of arsenic, trace organics, and coliform bacteria.

Navajo Nation EPA (Navajo EPA) estimates that up to 30% of the population on the Navajo Nation is not served by a public water system. This represents approximately 54,000 people. Residents of these homes either haul water from unregulated sources, such as livestock wells, springs, or private wells, or from regulated watering points. The number of unregulated water sources is not known, but is estimated to be in the low thousands.

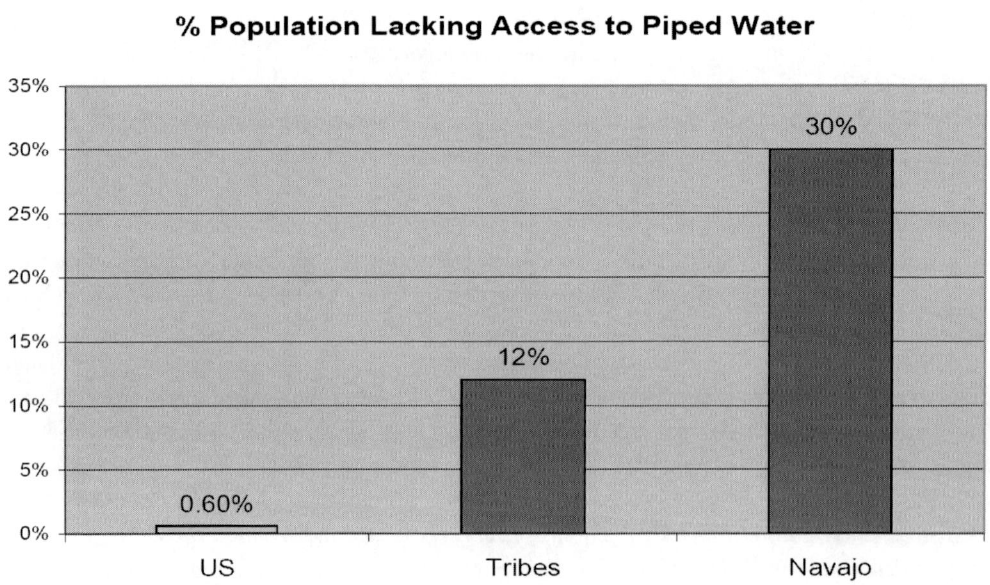

These unregulated sources generally do not fall under the jurisdiction of either USEPA or Navajo EPA, and are not routinely monitored to determine the quality of water being consumed. USEPA is concerned that some of these sources may be contaminated as a result of uranium mining activities. Data is not available at this time to determine the extent to which contamination is naturally occurring. The use of unregulated sources represents the greatest public health risk associated with drinking water for the Navajo Nation. Due to the potential acute and chronic health problems associated with drinking from unregulated sources, the Navajo EPA has consistently held that this practice is not sanctioned. Nevertheless, interviews with Navajo residents have shown that the practice continues and is widespread.

USEPA and the U.S. Army Corps of Engineers (USACE) surveyed unregulated water sources in the late 1990's under the Superfund Program, and found that 38 (17%) of the 226 unregulated water sources surveyed showed elevated (above the MCL) radionuclides in one-time grab samples. This sampling method is not a reliable technique for determining actual levels, but can serve as a starting point for further investigations. USEPA considered this information when compiling a prioritized list of 500 uranium mines, recently provided to Navajo EPA. Navajo EPA has identified at least two wells with elevated uranium levels that are currently in use as haul-water sources for one or more Navajo families. In addition, the impacts of uranium mining threaten the Hopi Reservation, which is surrounded by the Navajo Nation. Groundwater flow gradient information from the Navajo Tuba City Dump indicates that a contaminant plume is headed towards the drinking water source for the Moenkopi Villages.

In 2006 and 2007, Navajo Nation Environmental Protection Agency (NNEPA) and the Centers for Disease Control and Prevention (CDC) tested water from 199 unregulated water sources (livestock wells and springs) in order to determine the quality of unregulated water sources being used for drinking water by Navajo households. Water was analyzed for bacterial and chemical contaminants including uranium. Nine out of 199 (5%) unregulated wells tested exceeded the limit (30ug/L) for uranium in drinking water. The highest uranium level was 260 ug/L. Currently, CDC is working with US EPA, NNEPA and the Navajo Division of Health to implement a cross-sectional household survey in order to determine the extent of human exposure to uranium, arsenic and other chemical contaminants from consumption of unregulated drinking water and to assess the potential for health effects from that exposure. These investigations include data from non-mining regions of the Navajo Nation where groundwater contamination with uranium, arsenic and other minerals may naturally occur.

Action Plan

USEPA will use Superfund authority to take appropriate action at water sources contaminated by mine waste and found to be endangering human health. The USEPA Superfund Program, working with Navajo EPA and other partners, will begin a sampling effort in February 2008 to confirm potential contamination at 70 water sources in rural areas across the Navajo Nation identified as being at risk. Results from this sampling will be shared and discussed with Navajo Nation EPA, IHS, other Navajo agencies, New Mexico Rural Water Association, and the DiNEH Project to determine appropriate future actions. Sources with elevated uranium, but lacking evidence of mining impacts, may not be appropriate for action by EPA's Superfund Program due to Superfund's limitation on response to naturally occurring substances. In most cases future actions are anticipated to focus on public education and outreach to households using contaminated water.

USEPA has also made available $350,000 to Navajo EPA for a circuit rider program to provide technical assistance to public water systems and to families using unregulated water sources for FY08 and FY09. USEPA will provide laboratory services for the circuit rider to allow for limited monitoring of unregulated water sources. The circuit rider will use data from EPA's Superfund program and other sources to target homes thought to be using contaminated water. The Arizona Small Utilities Association and New Mexico Rural Water Association will assist with this effort. We expect that these personnel will be in the field in the second half of 2008, and continue throughout the five-year period of this plan.

Additionally, in FY08 USEPA Region 9 will allocate up to $400,000 of tribal source water funds, to be awarded as a contract, assistance agreement, or a combination of both, to the Navajo Nation. This funding will focus on efforts to protect current and future sources of drinking water used by the Tribe and its members.

The sources to be sampled by USEPA do not currently meet the Safe Drinking Water Act's definition of a public water system, which applies only to those systems that regularly serve an average of at least 25 people each day for at least 60 days per year or have at least 15 service connections.

Water Pumped by Windmill.

Limitations

The Indian Health Service has estimated the cost of providing alternative water supplies for eligible Indians consuming water from 41 unregulated and potentially contaminated water sources to be up to $65 million, including planning, design, and construction costs. This amount is not accounted for in the IHS budget, and will need to be recalculated after more reliable techniques are used to determine contamination levels. The Navajo Area Indian Health Service will expend nearly $19.5 million in FY08 for sanitation services, including providing access to clean water to eligible Indian homes.

The Navajo Nation will receive $34M over the five-year period of this plan from the Office of Surface Mining under Section 411(h)(1) and 411(h)(2) of the Surface Mining Control and Reclamation Act. The Navajo Nation may choose to use these funds to construct replacement water supplies in communities impacted by uranium mining; however, funds are not required to be used for this purpose.

ABANDONED URANIUM MINE ASSESSMENT AND CLEANUP

Agency: EPA

Background

Abandoned Uranium Mines: Mine operators extracted nearly four million tons of uranium ore from 1944 to 1986 across the 27,000 square-mile Navajo Nation. From 1994 to 2007, USEPA Region 9 conducted a screening-level investigation to identify approximately 1,100 mine features (portals, prospects, rim strips, pits, vertical shafts or waste piles) associated with AUMs throughout the Navajo Nation. These features represented 520 discrete AUMs. EPA also took several removal actions at high-priority sites, and has expended over $13 million on the investigation and removals combined. As partners, EPA has consulted with Navajo Nation Environmental Protection Agency (NNEPA) and relied on its local knowledge at all phases of the Navajo Nation-wide investigation and specifically during each of the removal actions.

The Navajo Abandoned Mine Lands Program (AML) used DOI Surface Mining Control Reclamation Act (SMCRA) funds to address an estimated 90% of the surface hazards posed by AUMs. Some of the mines continue to release contaminants into the environment and will require further cleanup or future O&M.

EPA is conducting a CERCLA removal action at the highest priority mine site, the Northeast Church Rock Mine. The Action Plan for Northeast Church Rock Mine is discussed separately below.

In August 2007, to assist the NNEPA with assessing the 520 AUM sites, EPA completed the Preliminary Site Ranking list, the project summary, the atlas, and the GIS database. This information will allow NNEPA to prioritize and categorize the 520 AUM sites into groups which might include removal, NPL, EPA or Navajo enforcement, Brownfields, and No Further Action.

Action Plan

USEPA will continue to work with NNEPA to decide on the highest priority sites and identify the appropriate course of action for each one. EPA will also continue to work with

Navajo Nation to provide resources to successfully manage our consultation and collaboration.

EPA and the Navajo Nation will work together to conduct a tiered assessment of abandoned mines, with more detailed assessments of those most likely to pose environmental or health problems. EPA has completed a screening-level assessment and risk-ranking of 520 known mines. In 2008, EPA will use this list and consult with the Navajo Nation to identify the highest-priority 200-250 mines for further evaluation. EPA plans to conduct more detailed on-site screening evaluations of 40-50 of these mine sites in 2008 and 2009, and would anticipate conducting future screening throughout the next several fiscal years until completed. EPA also plans to conduct removal assessments or preliminary assessments at approximately seven of the highest-priority mine sites in 2008 and 2009, and to continue to conduct such assessments in the following three years.

In deciding what sites to screen next, we would consider whether or not a mine has been reclaimed, whether a mine could impact drinking water sources, and other factors associated with risk and exposure. As part of the site screen, EPA and NNEPA will evaluate any available information, including data collected from the drinking water sampling effort that EPA and the Navajo agencies are currently conducting.

Based on the screening process, EPA will identify candidates for removal assessments or Preliminary Assessment/Site Inspections (PA/SI), or both, depending on the findings. Based on the screening process, many of the AUM sites may not pose a significant health risk and will require No Further Action.

In the past few years, NNEPA has screened 39 mine sites. Of these, 26 sites were classified as No Further Action, nine were referred to Navajo Abandoned Mine Lands, one was referred to Navajo Department of Justice, one was referred to BIA, one was referred to EPA for a removal assessment, and one was designated as EPA lead (Northeast Church Rock Mine).

In FY08 and FY09, EPA plans to conduct up to 50 on-site screening evaluations per year and to continue to conduct screens in the following three years. EPA currently provides grant funding for NNEPA to conduct up to 32 screens per year. EPA also plans to conduct up to seven preliminary assessments (PAs) or removal assessments in FY08 and FY09. Upon NNEPA's request, USEPA is ready to initiate CERCLA response actions at additional priority AUM sites that meet EPA's removal action criteria.

As a result of the screening process, PA/SI, and removal assessments, the sites will be categorized into groupings such as removal actions, NPL listing and subsequent EPA-led remedial actions, EPA or Navajo enforcement, Brownfields, and No Further Action. Based on the information in EPA's database on AUMs, EPA, under CERCLA, expects to address one priority AUM site in FY08, up to two of the highest-priority mine sites in FY09, and to continue such actions in the following three years.

Enforcement: High priority AUMs will be identified as candidates for further Superfund assessment and enforcement. EPA will continue to pursue our enforcement first policy. EPA is conducting a general search for Potentially Responsible Parties at all of the AUMs and will issue information request letters to several major viable corporations (responsible parties) that owned or operated multiple mines in 2008. EPA will continue to assist Navajo Department of Justice with Navajo-lead enforcement actions.

Costs: We expect to use our enforcement authority at AUMs that have viable PRPs. If there are no viable PRPs, then EPA will explore funding options as part of our budget request process to enable fund-lead response actions.

Assessment and Cleanup Cost Estimates: EPA has developed very preliminary estimates for the assessment and cleanup of the AUM sites. These costs were developed using cost estimating software and include general assumptions regarding the level of assessment and cleanup required.

Site Screens: 50 Site Screens per year will cost an estimated $100,000.

Assessment: the five Preliminary Assessments of aggregate AUM sites that NNEPA previously identified as priority areas have an estimated cost of $80,000.

Cleanup: the cleanup costs for an AUM range from several hundred thousand dollars to several millions of dollars. Until assessments and decision-making are complete, it is difficult to estimate with any confidence the scope of total cleanup costs.

To support our collaboration with Navajo Nation on this work, EPA's Superfund program is providing over $220,000 in additional grant funding to Navajo Nation for FY09, as compared to FY08, for site assessment and tribal response program funding to address AUMs. This funding includes an Inter-Personnel Agreement (IPA) from EPA to the Navajo Nation Superfund program.

The above cost estimates are very rough since so many critical cost-sensitive assumptions had to be made. The size of the Navajo Nation, the remoteness of many AUM sites, and the dearth of roads and infrastructure may add significant challenges and costs. EPA and the U.S. Army Corps of Engineers are currently refining these cost estimates and will incorporate refined estimates into the agencies ongoing budget processes.

Limitations

Legislation: Unlike the uranium mill tailings cleanup program, there is no specific legislation to address abandoned uranium mines. After the expiration of UMTRCA, neither UMTRCA nor CERCLA provides response authority for releases to soil from uranium mill sites. CERCLA is not ideally tailored to the AUM problem because of low population density in the Navajo Nation.

Transportation and Disposal: For mine sites that require remediation, options are to dispose of wastes in an onsite repository or offsite. Clean closure of these sites would require transporting the waste to a disposal site, which may be hundreds of miles away. There are potential health impacts from transporting the waste that must be weighed against leaving the waste on site in a suitably constructed repository. Further, it will be difficultto find an off-site disposal location that is capable of accepting the waste. Transportation and disposal arrangements need to be discussed in advance with the Navajo Nation as considerations in selecting a suitable remedy.

Cleanup of the Northeast Church Rock (NECR) Mine Site

Agency: EPA

Background

Located near Gallup, New Mexico, NECR is the highest priority on the AUM ranking list. The mine adjoins the United Nuclear Corporation uranium mill site, a National Priorities List cleanup project that USEPA Region 6 manages jointly with the Nuclear Regulatory Commission. The mine is mostly on Navajo tribal trust land, while the mill is on private fee land. At the request of the Navajo Nation, USEPA Region 9 is using Superfund removal authority to investigate and clean up the NECR mine site, in coordination with the NPL site cleanup. USEPA issued an order to the responsible party, United Nuclear Corp (UNC) which is now a subsidiary of GE, to investigate and assess risk at the residential area adjacent to the mine. In summer 2007, EPA removed 6,500 cubic yards of radium-contaminated soils around these residences to address the highest risk areas. EPA spent $990,000 on the excavation, and required GE to dispose of the soils at an additional cost of $1.3 million.

Action Plan

To determine the remedy for the long-term threats from the NECR Mine, EPA has developed an engineering evaluation/cost analysis (EE/CA), which includes a preferred cleanup alternative. In May 2008, Navajo Nation requested that EPA hold the release of the EE/CA until September 2008 to allow for negotiations between Navajo Nation and General Electric (GE).

EPA is poised to release the EE/CA in September 2008 for public comment and select the soil remedy in December 2008. Depending on the outcome of the negotiations between the Navajo Nation and GE, EPA may order GE to implement the selected remedy, which will cleanup the balance of the site by addressing additional contaminated soils with an estimated volume of up to 870,000 cubic yards. Cleanup is expected to take up to three years.

Costs: EPA anticipates that GE will conduct and fund the work at NECR over the next several years under an enforceable agreement or a Unilateral Administrative Order issued by EPA. If GE does not comply, EPA may spend some Superfund money for the Remedial Design and cost recovery, while working to enforce against GE.

Long-Term Management of Former Uranium Milling Sites

Agency: DOE and NRC

Background

Title I of the Uranium Mill Tailings Radiation Control Act (UMTRCA) originally required the cleanup of 22 inactive uranium milling sites and the properties in their vicinity, as well as the properties in the vicinity of the Tennessee Valley Authority Edgemont uranium milling site. In 2001, Congress also required DOE to remediate the Moab uranium milling site under UMTRCA. DOE's authority for surface cleanup expired in September 1998 (except for its current authority to remediate Moab); there is no end date for maintenance of existing groundwater remediation activities. Four of the original uranium milling sites (Shiprock, NM; Tuba City, AZ; Monument Valley, AZ; and Mexican Hat, UT) were located within the Navajo Nation's lands.

In 1983, DOE and the Navajo Nation entered into a cooperative agreement to facilitate the surface remedial action of the four sites. Funds provided by DOE to the Navajo Nation for the agreement were directed to the Navajo Nation's Division of Natural Resources. The National Environmental Policy Act (NEPA) compliance documentation which includes the Environmental Assessments as well as the Remedial Action Plans were developed with the participation of the Navajo Nation. The characterization of surface and subsurface radiological contamination involved a phased approach that included an aerial assessment, a mobile scan, and on site sampling to delineate the extent of contamination according to specific removal criteria set by the U.S. EPA. Over $140 million was spent on the surface remediation of the four processing sites and construction of three disposal cells. That agreement expired in 1998 along with the surface cleanup authority, and a new cooperative agreement was initiated for groundwater cleanup that remains in place.

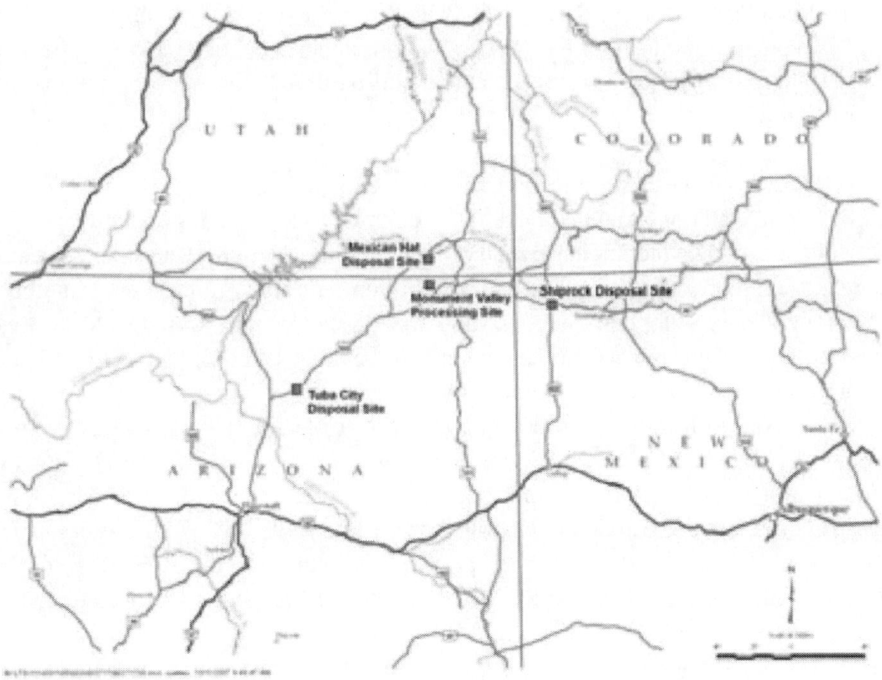

DOE's Office of Legacy Management, in consultation with the Navajo Nation has been addressing groundwater contamination at the sites. The Mexican Hat site groundwater contamination is confined and the hydraulic conditions prevent the future use of any groundwater within the area, therefore no further remedial action is warranted at that site.

Action Plan

As required by UMTRCA, DOE has worked closely with the Navajo Nation Division of Natural Resources, and with other federal and state agencies in the clean up of uranium milling sites.

Milling Site Actions

- DOE will continue to monitor the three disposal cells to ensure they remain effective in protecting human health and the environment. DOE's long term surveillance and maintenance duties will continue at the three disposal cells for the foreseeable future.
- DOE funds maintenance of groundwater remediation activities and long- term surveillance and maintenance at the four Navajo Nation sites at an estimated $3.2 million in 2008. The groundwater compliance strategies are reviewed annually with the Navajo Nation to track progress towards cleanup standards in the groundwater plumes. Based on our knowledge of the sites, DOE does not believe that additional funding would significantly accelerate the groundwater cleanup at the three processing sites where ongoing pump and treat remediation activities are being maintained. DOE will continue to work with the Navajo Nation using the currently approved plans to complete our groundwater remediation responsibilities.
- DOE will continue to fund the Navajo Nation, under the Cooperative Agreement, to assure they have the resources to review and participate as we continue these long term actions.

Note: A letter from the Navajo Department of Justice, dated March 24, 2008, reiterated the desire to "remove all contaminated materials from all UMTRCA sites … with final disposition outside of Navajo Indian country." The Navajo Nation request to relocate the disposal cells would: be costly (potentially $1 billion or more); create opportunities for adverse health impacts – to both workers and the public; require a state (New Mexico, Arizona, Colorado, or Utah) to accept the contents of the disposal cells; and, not eliminate the need for current groundwater remediation efforts.

The Navajo Nation has also raised a question about placing a liner under the disposal cells. Lining the disposal cells would not have significantly enhanced the groundwater cleanup status either in the past or if the disposal cells were relocated to lined cells in the future because groundwater contamination is primarily due to the historical milling activities at the sites and was already well established prior to initial cleanup by DOE. None of the 22 UMTRA Title I disposal cells are lined. The Department has offered to consult with the Navajo Nation on these issues.

Federal agencies (DOE, EPA, IHS, BIA, DOI, and NRC) met with representatives from the Navajo Nation, the Hopi Tribe and EPNG in Tuba City February 12-13, 2008, to continue efforts to exchange technical information and present scientific findings related to the characterization and remediation of the Tuba City dump, the Highway 160 site, and the groundwater treatment system at the DOE Tuba City disposal cell. During the February meeting, it was agreed that all parties will continue efforts to formally document and share the information regarding the three sites. DOE coordinated the distribution in digital form, to all parties, of all known information related to the Highway 160 site and the former uranium milling site on May 29, 2008. The affected federal agencies, Navajo Nation and Hopi Tribe, will reconvene in August to collectively assess that combined set of data and information.

Other Related Actions

DOE will continue to work with U.S. EPA to provide information on historical vicinity property surveys data to determine the scope of other contaminated properties on the Navajo Nation's lands.

DOE will work with BIA and EPA in interpreting the groundwater data they have collected at the Tuba City Dump site as they have requested.

In accordance with UMTRCA Title II, DOE anticipates it will eventually receive the Church Rock uranium milling site after its owner completes reclamation and NRC approves termination of its radioactive materials license, if the State does not exercise its option. DOE will continue to work with the Navajo Nation regarding the reclamation and transfer of this site into DOE long term management to ensure protection of human health and the environment.

5 Year Plan Cost Table

Site	2008	2009	2010	2011	2012
Mexican Hat	Annual inspection	Annual inspection	Annual inspection	Annual inspection	Annual inspection
Monument Valley	Biannual inspection Groundwater monitoring	Biannual inspection Groundwater monitoring	Biannual inspection Groundwater monitoring	Biannual inspection Groundwater monitoring	Biannual inspection Groundwater monitoring
Shiprock	Biannual Inspection Groundwater monitoring Maintain groundwater treatment	Biannual Inspection Groundwater monitoring Maintain groundwater treatment	Biannual Inspection Groundwater monitoring Maintain groundwater treatment	Biannual Inspection Groundwater monitoring Maintain groundwater treatment	Biannual Inspection Groundwater monitoring Maintain groundwater treatment
Tuba City	Biannual Inspection Groundwater monitoring Maintain groundwater treatment	Biannual Inspection Groundwater monitoring Maintain groundwater treatment	Biannual Inspection Groundwater monitoring Maintain groundwater treatment	Biannual Inspection Groundwater monitoring Maintain groundwater treatment	Biannual Inspection Groundwater monitoring Maintain groundwater treatment
Total Cost	$4,940,000	$3,178,000	TBD	TBD	TBD

Note: Funding in 2010-2012 has not been determined, but is projected to be consistent with 2008 and 2009 funding levels.

Limitations

The Legacy Management program appears secure into the foreseeable future to maintain the disposal cells and meet the groundwater plans as agreed to by DOE and the Navajo Nation.

Characterization and Remediation of Highway 160 Site

Agency: DOE and EPA

Background

The Highway 160 Site is located approximately four miles northeast of Tuba City, Arizona, on the north side of Highway 160. The Department of Energy maintains a disposal cell and operates a groundwater treatment system on the south side of Highway 160. The Department of Energy conducted characterization and remediation of the former uranium mill site (south of Highway 160) under the Uranium Mill Tailings Radiation Control Act (UMTRCA) in the 1980s and 1990s. A radiation survey conducted by DOE in 1980 as part of the Vicinity Properties program, on the north side Highway 160 Site of Highway 160 indicated no contamination in this area. This is probably because there was sufficient soil coverage to mask any gamma radiation from the material that has since been uncovered by wind erosion.

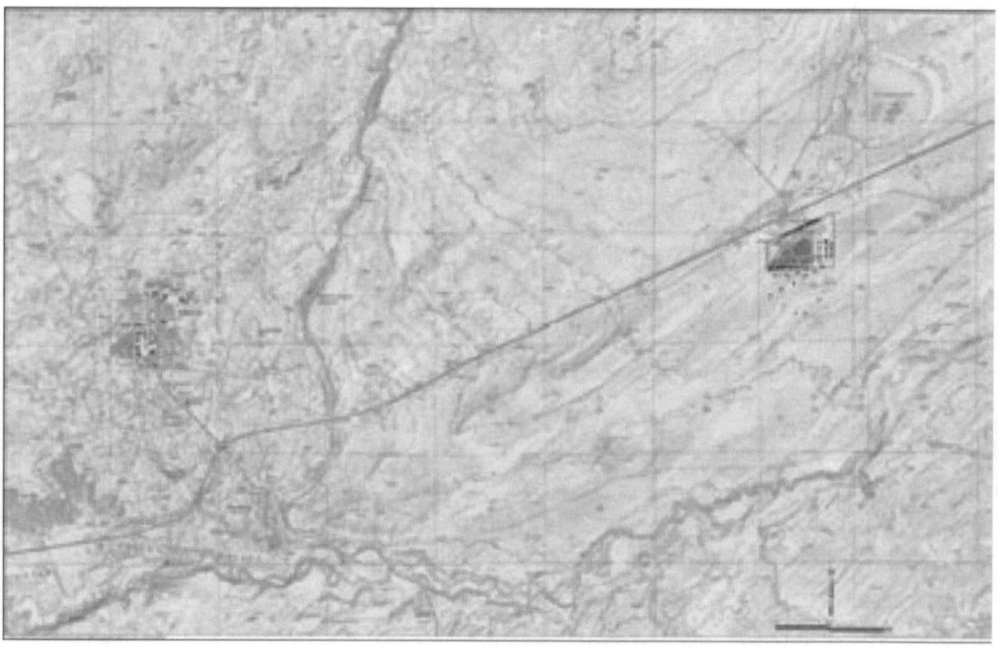

In a December 1, 2003 letter, the Navajo Nation notified DOE that materials suspected to be originating from the former Rare Metals uranium milling site were found at the Highway 160 site. At that time a response was sent indicating that the DOE authority for surface remediation under UMTRCA expired in 1998.

In May 2004, EPA conducted a radiation survey of the Highway 160 Site. In May 2007, EPA offered to have an On-Scene Coordinator revisit the site based on the proximity of a new residence, further evidence asserted in a Navajo EPA report, and EPA's observations during a site visit. In response, Navajo EPA/NDOJ expressed their desire to retain the lead for the site and to defer EPA follow-up until further notice.

El Paso Natural Gas (EPNG), a former operator of the Tuba City inactive uranium milling site, and Navajo Nation EPA recently entered into a cooperative agreement to characterize the site and conduct interim remedial actions. Characterization activities and interim actions (a fence and stabilization of soil using a fixative) were performed by EPNG. While the origin or exact nature of the contamination at this site is not clear EPNG site characterization data does indicate some radioactive debris. EPNG has also sued the Federal Government, including DOE, NRC, EPA, DOI, BIA, HHS, IHS, and the Department of Defense, seeking that these agencies clean up the Highway 160 site and the Tuba City dump.

Action Plan

DOE will work with the other affected Federal agencies, including the BIA, USGS, and EPA as appropriate, the Navajo Nation, and EPNG to reach a consensus on the best approach for characterization and any required cleanup based on characterization of the site.

Federal agencies (DOE, EPA, IHS, BIA, DOI, and NRC) met with representatives from the Navajo Nation, the Hopi Tribe and EPNG in Tuba City February 12-13, 2008, to continue efforts to exchange technical information and present scientific findings related to the characterization and remediation of the Tuba City dump, the Highway 160 site, and the ground water treatment system at the DOE Tuba City disposal cell. During the February meeting, it was agreed that all parties will continue efforts to formally document and share the information regarding the three sites. DOE coordinated the distribution in digital form, to all parties, of all known information related to the Highway 160 site and the former uranium mill site on May 29, 2008. The affected federal agencies, Navajo Nation and the Hopi Tribe, will reconvene in August to collectively assess that combined set of data and information.

There are currently three authorized courses of action related to additional characterization and remediation of the Highway 160 site: (1) EPA cleanup of the site as a Superfund enforcement action; (2) continued action by EPNG; (3) some combination of the two approaches.

If EPA determines that an Imminent and Substantial Endangerment exists at this site, EPA can use Superfund enforcement authority to require the potential responsible party to conduct a response action. DOE will cooperate with that effort as requested by the parties, as appropriate. If El Paso Natural Gas works to affect a cleanup of this site, DOE will work with them to consider disposal of the contaminated materials at the DOE- operated Cheney Disposal Cell near Grand Junction, CO.

Based on existing site knowledge, DOE estimates it would take several years to complete the work at the Highway 160 site. DOE has provided support for the preparation of a cost estimate for the remaining cleanup work based on the EPNG characterization report (surface radiation levels and magnetometer readings). The analysis considered a work scope that would include completing site characterization efforts, preparing a remedial action plan, obtaining regulatory approval for the plan and executing the work, obtaining approval for disposal of the material that was removed, and completing other related cleanup actions, as appropriate, at the Highway 160 site. The cost estimate would require additional characterization data that provides information on the chemical species and the depth of the contamination.

Regardless of which entity conducts the site characterization and remedial action, DOE will work with the affected federal agencies and the Navajo Nation and Hopi Tribe to locate additional groundwater monitoring wells in the vicinity of the former uranium milling site, as necessary, to improve our understanding of groundwater flow in the area.

Limitations

DOE does not have the authority to perform surface characterization or remediation of the Highway 160 site as our authority under UMTRCA for these activities expired in 1998.

Cleanup of Tuba City Dump

Agency: Bureau of Indian Affairs (BIA), Central Office

Background

The Tuba City Dump (Dump) was used for more than 50 years as an open, uncontrolled dump receiving solid waste from local communities. For some years, the BIA maintained the Dump by covering trash and burying waste. The Dump was closed administratively by BIA in August 1997. It was the principal open dump site in the Tuba City/Moenkopi area during this period. Information on wastes disposed of at the site is limited as the site was unmanned and unsecured. The Dump occupies land both on the Hopi Reservation and the Navajo Nation. The BIA, along with the Environmental Protection Agency (EPA), and Indian Health Service (IHS), are working with the Hopi and Navajo tribes to address environmental concerns at the site. These concerns include persistent elevated levels of uranium, vanadium, radium 226/228, gross alpha and beta activity, and other metals in the shallow groundwater at the site. Deeper groundwater and nearby drinking water supply wells have, to date, not exhibited elevated levels of the above constituents.

Approximately four miles northeast of the Dump is the Tuba City inactive uranium milling site (mill) that is currently being managed by the Department of Energy (DOE) under the Uranium Mill Tailings Radiation Contract Act (UMTRACA) program. The Navajo and Hopi tribes have long maintained that waste from the mill was improperly disposed of at the Dump. A U.S. Geological Survey (USGS) study, dated October 2006, suggests that the radionuclides in the shallow groundwater may be from a different formation than the native rock at the site.

Since 1999, the BIA has been conducting assessment activities of uranium contamination of shallow and deep groundwater, springs, contaminant migration pathways, sources, and receptors for the purpose of formulating a final closure plan. The springs are used by the Hopi Tribe for religious and ceremonial purposes. Groundwater monitoring data indicates that uranium containing materials have been disposed of at the site creating a uranium plume in the shallow groundwater which extends to the west and southwest of the site. Additionally, groundwater monitoring data shows that the uranium contamination of the shallow groundwater is present up gradient of the site indicating that the contamination is not limited to the Dump. Other sources, up gradient to the Dump, appear to be contributing to shallow groundwater contamination; however, the exact source of up gradient contaminant contribution is unknown at this time.

Landfill characterization studies have not detected contamination in landfill materials at levels that constitute hazardous waste or radioactive waste. Soil vapor studies detected no hazardous waste constituents and determined that methane was not migrating from the landfill. Both tribes want "clean closure" (removal of all waste).

Action Plan - Overview

The five federal agencies worked together with the Navajo and Hopi tribes in early 2008 to assess whether interim actions are needed prior to selection of a final remedy for the site. By August 2008, the BIA will complete an assessment of the need and feasibility of conducting an interim measure to prevent contamination of nearby Hopi water supplies. If an imminent threat to water supplies is identified, the agencies will determine the most appropriate authorities to achieve an interim remedy. These authorities might include a Superfund response or enforcement action.

For 2007 and FY 2008 $916,000 will be used to perform additional studies needed to determine the movement of contaminated groundwater, and potential threat to drinking water supply wells, irrigation water, springs, and seeps. These studies are also necessary in order to determine the need for, and scope of remediation. The following tasks include a break down of the $916,000.

Task 1: RI/FS Work Plan

The BIA has contracted with SECOR International to prepare the RI/FS work plan with the option for limited additional studies. The purpose of the RI/FS work plan is to identify the tasks and establish a schedule to fully investigate contamination impacts associated with the Dump and to develop cleanup options. At the time the contract was signed in October 2007, BIA estimated the costs of the RI/FS work plan and additional limited studies to be $204,000. Since the contract was let, results of other studies and the conceptual site model (developed by SECOR) suggest the need for additional studies and interim actions to mitigate potential health risks.

The development of the RI/FS study may identify other parties responsible for site closure. The RI/FS work plan will be completed by October 2008.

Task 2: Limited Additional Studies

Based on the conceptual site model developed by BIA's contractor and on information recently provided by the Navajo Nation, limited additional studies and actions are necessary to assess and evaluate the potential imminent threat or risk to public health and the environment posed by the Dump. These studies are an extension of those anticipated in the limited additional studies contracted in Task 1 above and will be completed by August 2008. These studies are intended to support the RI/FS work plan and will evaluate the location and concentration of the contaminant plume migrating from the Dump and to assess whether that plume threatens existing domestic and irrigation water supply systems. Since this information was provided to BIA at a meeting in Tuba City, Arizona, on December 6, 2007, several conference calls have been initiated with members of the USGS, BIA, IHS, Hopi Tribe, Navajo Nation and DOI, as well as the BIA contractor, SECOR International, the Navajo Nation's contractor, William Walker and Associates, and the Hopi Tribe's contractor Daniel B. Stephens and Associates. The scope of additional studies was finalized on March 31, 2008, and is outlined below:

- Well Installation—10 shallow groundwater wells and two deep groundwater monitoring wells were initially proposed. As field work progressed, the technical team determined that two additional wells (one shallow and one deep) were necessary. The wells were proposed to assess water quality migrating from potential up-gradient sources northeast of the dump and in Pasture Canyon and to assess the lateral and vertical extent of potential impacts from the dump. Two well clusters consisting of shallow and deep wells were installed on the west side of Pasture Canyon between the dump and the Moenkopi water supply wells. These wells are intended to be used as sentinel wells for migration of potential contaminants toward the water supply wells. The well boreholes will be continuously logged. The deep wells will be drilled with hollow stem auger, while the two deep wells will be advanced using air rotary to penetrate bedrock. Five of the shallow wells will be completed with 4- inch PVC casing for potential use as extraction wells, while the remaining nine wells will be completed as 2-inch wells. Well installation was initiated on April 1, 2008, and was completed on May 8, 2008. Total costs for well installation are expected to be on the order of $193,000.
- Baseline Groundwater Monitoring Event—The baseline monitoring event will be conducted on all 52 groundwater monitoring wells, three supply wells, seven springs, the irrigation pipeline water and the Pasture Canyon Reservoir. In addition, discrete sampling will be performed in up to 12 groundwater wells to assess water quality in near surface alluvium and bedrock. Each sample will be analyzed for a comprehensive suite of analytes to differentiate differences in background water quality and leachate from the dump. Groundwater sampling was initiated on May 5, 2008, and was completed at the end of May. Total costs for the entire baseline monitoring event are $262,000.
- Aquifer Testing—If an imminent threat to water supplies is identified, a groundwater pump test will be conducted to evaluate design parameters for an interim remedy. The fees for this task include installation of an extraction well, two observation wells, a 72-hour pump test, and slug testing of up to 19 wells. Costs for this scope of work are $65,000.

- Permitting, Internet Data Repository and Other Studies—In addition to the studies above, BIA will require the contractor to participate in public meetings, develop a web repository for project reports and documentation, obtain permits, access agreements, encroachment permits for additional studies, prepare meeting minutes, and other tasks. The estimated costs for this task are $111,000. An additional $60,000 is planned for subsequent years.
- Aerial Survey and 5-foot Contour Map—BIA plans to develop a five-foot contour map of the landfill and area of potential impact. This area comprises approximately 2,000 acres. This map will be used to evaluate the potential connection between the landfill and down-gradient drinking water and irrigation springs. The cost to perform the aerial survey and to develop contour maps is $17,000.
- USGS will evaluate the hydrologic connection between the area of the Dump and down gradient water supply resources of Pasture Canyon and Upper Moenkopi Wash. This study is necessary to determine the possibility and likelihood of these resources to be impacted from uranium in the shallow groundwater emanating from the Dump, and ultimately for justification of an interim groundwater action in the event that hydrologic connectivity between the source(s) and receptors is established. The USGS study and Hydrologic Mapping contract will cost $64,000.

Task 3: Interim Remedial Measures

If an imminent threat to water supplies is identified, the agencies will determine the most appropriate authorities to achieve an interim remedy.

Task 4: Groundwater Monitoring

It is expected that groundwater monitoring will be conducted on a semi-annual basis throughout the remedial investigation process. Further, groundwater monitoring is planned to occur until 2042, 30-year post-closure monitoring, unless the remedy selection indicates a reduced post-closure monitoring period. It is assumed that up to 10 wells will be monitored for key analytes. The costs for each groundwater monitoring event are estimated at $20,000 per event ($40,000 per year) with a larger suite of analytes monitored during the first semester of the fifth year.

Task 5: RI/FS

Based on the outcome of the additional limited studies and the RI/FS work plan, it may be necessary to assess:

- Source(s) of elevated concentrations of contaminants in up-gradient wells using geophysical and radiological survey methods as well as groundwater wells;
- Contaminant migration through bedrock by installing two to three additional deep wells to evaluate potential contaminant migration south and down-gradient of the Dump;
- Groundwater migration characteristics using tracer tests; and
- Source of elevated contaminants in the dump using larger diameter borings or trenches.

In addition, feasibility studies may be necessary to evaluate the feasibility of various groundwater treatment options and landfill closure alternatives.

A baseline risk assessment would be performed as part of the RI/FS, if necessary. The purpose of the baseline risk assessment would be to "characterize the current and potential threats to human health and the environment that may be posed by contaminants migrating to ground water or surface water, releasing to air, leaching through soil, remaining in the soil, and bio-accumulating in the food chain" (40 CFR Section 300.430(d)(4)). The health risk assessment would evaluate chemicals of concern, exposure pathways, and potential receptors (both environmental and human). The baseline risk assessment would provide the BIA and other stakeholders with an understanding of the actual and potential risks to human health and the environment posed by the site and any uncertainties associated with the assessment. This information would be necessary to determine whether a potential threat to human health or the environment exists and would aid in assessing the need for and type of remedial actions or closure criteria that would be applied at the site.

Implementation of this RI/FS work is planned to occur between November 2008 and December 2009.

Task 6: Stakeholder and Other Government Agency Costs

The BIA is currently retaining the expertise of stakeholders and other government agencies in assessing and evaluating environmental impacts and closure alternatives for this site.

Task 7: Remedy Selection and Closure Plan Design

Using the CERCLA alternatives analysis methodology, a preferred remedy will be selected from the RI/FS study based on technical and stakeholder input. This remedy will involve the implementation of multiple alternatives to address all potential contaminants of concern, and completed routes of exposure and receptors. The remedy may include groundwater remediation. This selection process will involve a Record of Decision (ROD) concurred on by EPA. The remedy selection process is planned to occur during the first half of 2010.

Task 8: Site Closure

The closure plan design will include engineering design, planning, management, community participation for closure and post-closure activities at the site. A site closure plan will be developed as a result of the ROD. Remedial activities may include the complete or partial excavation of waste, groundwater treatment, waste disposal including transportation, hydro-seeding and restoration, dewatering and leachate evaporation ponds, construction of repositories, etc. Once a remedy is selected, the closure and post-closure costs will be estimated. Closure activities are planned to occur between July 2010 and December 2012.

* BIA's action plan for the Tuba City Dump cannot be finalize until site testing data/results are received. BIA expects testing results by August/September.

HEALTH ASSESSMENT

Agency: Indian Health Service

Background

The Indian Health Service (IHS) has provided healthcare to Navajos (approximately 240,000 users presently) since 1955. The Area Office responsible for administering programs on Navajo lands is the Navajo Area Indian Health Service (NAIHS), headquartered in Window Rock, Arizona.

Extensive information on the health status of individual Navajos is available in patient medical records and in patient databases maintained by the IHS.

Much is known about the health effects of occupational and other exposures to humans from uranium and its breakdown products from articles in the scientific literature. Most of what is known about Navajos relates to occupational exposure.

The Navajo Nation believes that further epidemiologic studies would be of help to the Navajo people as they are concerned about risks to individuals exposed to uranium and its breakdown products, as well as other possible associated contaminants, left behind on Navajo lands from past mining and milling efforts.

The IHS has worked in the past with the Navajo Nation and individual eligible Indians seeking health care services at DHHS/NAIHS facilities on issues regarding exposure to uranium. In the last decade DHHS and NAIHS have invested an estimated $3.8 million providing specific medical screening examinations and patient evaluation reports to former uranium miners, former millers, and "downwinders" from Nevada test bomb explosions in earlier decades. This amount also includes costs for providing patient and community education on the hazards of uranium exposure, providing individual eligible Indians or their legal descendants with information on their specific health conditions and planning and administrative responsibilities of the agency associated with providing services to these individuals.

Action Plan

The IHS will continue current efforts and proposes new efforts for describing the health effects of contamination of surface lands and water on the Navajo people as discussed in the accompanying table. At this time, IHS does not plan to seek additional authorities nor additional funding to complete this action plan.

Objectives	Feb.'08	Aug.'08	Jan.'09	July'09	Jan.'10	July'10	Jan.'11	July'11	Jan.'12	July.'12
1. Diagnose and treat known health conditions in eligible Indians.	X Continue	X	X	X	X	X	X	X	X	X
2. Radiation exposure, screening, and education Program Grant Proposal/ Submission for 2009 to DHHS-HRSA.		X								
3. In-kind contribution support to Univ. of New Mexico Health Sciences Center "Navajo Uranium Assessment and Kidney Health Project" in northwest New Mexico Navajo communities funded by DHHS- NIH.	X Ongoing	X	X	X	X	X	X	X		
4. Meet with Navajo Nation Div. of Health Epi Center staff to share information and discuss potential future collaborations consistent with Federal Agency action plans and Navajo Nation health concerns.	X Ongoing	X	X	X	X	X	X	X	X	X
5. Participate in February 2008 onsite Technical Data Conference in Tuba City, AZ, discussing mill site contamination data/Tuba City dump and water contamination issues with Navajo Nation (NN), NRC, DOE, BIA, EPA.	X									
6. Review existing IHS patient databases and University of New Mexico (UNM) Tumor Registry databases with UNM staff and NN-EPI staff to develop plans for improved cancer case surveillance.		X Complete								

Objectives	Feb.'08	Aug.'08	Jan.'09	July'09	Jan.'10	July'10	Jan.'11	July'11	Jan.'12	July.'12
7. Review water contamination data from CDC (Health Studies Branch, Div. of Environmental Hazards and Health Effects, National Center for Env Health) with Navajo Nation, UNM, and EPA for potential future health studies.			X Complete							
8. Examine previous case-control study work and data from DOE, BIA, CDC, NRC, NN and UNM, and develop plans as appropriate to assess cancer and other health condition prevalences for populations near abandoned mill/mine sites including, if necessary, potential case-control study designs.			X Complete							

SUMMARY OF NRC ACTIONS

Agency: Nuclear Regulatory Commission

Background

- The U.S. Nuclear Regulatory Commission's (NRC's) regulatory program, based on the Uranium Mill Tailings Radiation Control Act of 1978 (UMTRCA), governs the remediation of legacy sites and licensing new uranium recovery (UR) facilities.
- During the 1980s after issuance of UMTRCA, the NRC implemented many regulatory improvements to avoid the problems from uranium milling operations conducted in the 1950s and 1960s. Some of these are:
    - Implementation of below grade disposal of mill tailings.
    - Use of liners and groundwater monitoring for tailings impoundments to prevent groundwater contamination.
    - Financial surety requirements to reduce the risk of problems caused by a licensee's bankruptcy.
    - Reclamation and decommissioning requirements.
- The NRC's UR program regulatory framework is designed to adequately protect public health, safety, and the environment.

NRC Actions

1. *UNC Church Rock Mill Site:* The NRC will continue to work with the Navajo Nation on issues related to the clean up of this old mill site. NRC staff has previously worked with the Navajo Environmental Protection Agency (EPA) on this action and will now also work with the Navajo Department of Justice (DOJ) (Mr. David Taylor) as requested by the Navajo Nation.
2. *New UR Applications:* The NRC will consult with the Navajo Nation, via the Navajo DOJ, for new UR license applications that have potential cultural and resource impacts on the Navajo Nation.
3. *U.S. Department of Energy (DOE) Title I Uranium Mill Tailings Sites:* For the four Title I mill sites located on Navajo lands (Mexican Hat, Monument Valley, Tuba City, and Shiprock), the NRC will continue to consult with the Navajo Nation and concur on DOE's remedial plans as required by NRC regulation 10 CFR 40.27. The NRC staff has previously interacted with the Navajo Nation (Navajo Environmental Protection Agency and Division of Natural Resources – Uranium Mill Tailings Remedial Act Department) on the review of Groundwater Compliance Action Plans such as the Shiprock site. In addition, as requested by the Navajo Nation, the NRC staff will also consult with the Navajo DOJ.
4. *Tuba City Contaminated Properties:* Under Title I of UMTRCA, the contaminated sites near the Tuba City mill site would have been considered vicinity properties if DOE determined, in consultation with NRC, that the contaminated material was

derived from the Tuba City mill site. NRC's role in the remediation of these vicinity properties was limited to concurrence in DOE's selection, performance, and completion of remediation of the vicinity properties. DOE would not obtain an NRC license to perform the remediation. However, the Tuba City mill site is under the general license in 10 CFR 40.27.

The purpose of the general license in 10 CFR 40.27 is to ensure that tailings disposal sites are cared for in such a manner to protect public health, safety, and the environment once the remedial actions are completed.

5. NRC staff members met with the tribes and held public meetings in Gallup, New Mexico, in September 2007, to explain the regulatory approach that NRC intends to follow for the review of proposed uranium mills. This meeting included a number of Navajo participants and NRC's presentation was translated into the Navajo language. The NRC has also conducted other meetings on this subject in Casper, Wyoming, and Albuquerque, New Mexico.

The Commission has provided the staff with adequate resources to perform safety reviews for new UR applications and maintain outreach with the Navajo Nation for FY 2008 and FY 2009.

In: Uranium Contamination in the Navajo Nation
Editor: Eugene Stepp
ISBN: 978-1-63321-678-5

Chapter 3

FEDERAL ACTIONS TO ADDRESS IMPACTS OF URANIUM CONTAMINATION IN THE NAVAJO NATION: FIVE-YEAR PLAN SUMMARY REPORT*

Bureau of Indian Affairs, Department of Energy, Nuclear Regulatory Commission, Environmental Protection Agency, and Indian Health Service

EXECUTIVE SUMMARY

The Navajo Nation encompasses more than 27,000 square miles, spread between three states in the Four Corners area. The unique geology of the region makes the Navajo Nation rich in uranium, a radioactive ore in high demand after the development of atomic power and weapons at the close of World War II in the 1940s. Approximately four million tons of uranium ore were extracted during mining operations within the Navajo Nation from 1944 to 1986. Many Navajo people worked the mines, often living and raising families in close proximity to the mines and mills.

Uranium mining and milling activities no longer occur on Navajo lands, but the legacy of these activities remains; including more than 500 abandoned uranium mine claims with thousands of mine features such as pits, trenches, holes, etc., and homes and drinking water sources with elevated levels of uranium, radium and other radionuclides. Uranium and other elements (selenium, arsenic, etc.) associated with mine and mill sites also occur naturally at elevated levels in rock, soil, surface water, and groundwater across the Navajo Nation and the broader Four Corners region. Health effects as a result of exposure to these elements can include lung cancer, bone cancer, and impaired kidney function.

* This is an edited, reformatted and augmented version of a report issued January 2013.

Historical Uranium Mining on the Navajo Nation - King Tutt Point Mine.

In October 2007, at the request of the U.S. House Committee on Oversight and Government Reform, the United States Environmental Protection Agency (EPA), along with the Bureau of Indian Affairs (BIA), the Nuclear Regulatory Commission (NRC), the Department of Energy (DOE), and the Indian Health Service (IHS) developed a coordinated Five-Year Plan to address uranium contamination in consultation with Navajo Nation. The Five-Year Plan was the first coordinated approach by the agencies, and outlined a strategy for gaining a better understanding of the scope of the problem and for addressing the greatest risks first. The following objectives were developed as part of the Five-Year Plan:

1. Assessment and cleanup of contaminated structures
2. Assessment of contaminated water sources, and provision of alternative water supplies
3. Assessment of abandoned uranium mines, with detailed assessments of those most likely to pose environmental or health problems
4. Cleanup of the Northeast Church Rock mine site and additional high-priority abandoned mine sites
5. Cleanup of the Tuba City Highway 160 site
6. Cleanup of the Tuba City Dump
7. Remediation of groundwater contamination at three former mill sites
8. Case control studies of health risks faced by individuals residing near mill sites or abandoned mine sites

This report is provided in response to a letter dated October 18, 2012 from six members of Congress requesting that the EPA, BIA, NRC, DOE, and IHS provide a summary of work completed toward the goals outlined in the Five-Year Plan. The members requested that the report include:

1. The extent to which the federal agencies have been successful and effective in accomplishing the cleanup and assessment milestones established in the Five-Year Plan.
2. A description of the results of the assessment and characterization efforts, including a discussion of changes to the initial understanding of the scope of the contamination problems over the past five years.
3. A discussion on how effective the collaboration among the federal and other agencies involved has been, and any ways collaboration and information sharing could be further improved during implementation of the next Five-Year Plan.
4. A preliminary discussion of the remaining immediate and longer-term steps that need to be taken to address the uranium contamination in and around the Navajo Nation.
5. A discussion of whether past allocated resources were sufficient to accomplish the milestones outline in the Five-Year Plan and anticipated future funding needs for additional remediation and public health efforts, including a discussion of the role of non- appropriated funds, such as those from PRPs, during the first five years and for future efforts.

1. Progress toward Milestones of the Five-Year Plan

Significant progress was made on all eight goals of the plan as described in detail below. The agencies met the overarching goal of protecting Navajo residents from urgent risks first by addressing contaminated structures, water supplies, and mills and mines with the highest levels of radiation. There is also an improved understanding of the scope of mine-impacted and naturally occurring uranium on the reservation. Uranium mining has resulted in elevated uranium levels at over 400 locations on the Navajo Nation, which will take additional time to address. All agencies are committed to continue working with the Navajo Nation to understand and address the health and environmental risks and to find long-term solutions to the remaining uranium issues on Navajo lands.

The agencies completed the following actions:

OBJECTIVE 1. *Assessment and cleanup of contaminated structures:* The EPA and Navajo Nation EPA (NNEPA) exceeded the Plan goal of assessing 500 homes and other structures that were likely to be contaminated with radiation at elevated levels. Together, the EPA and NNEPA surveyed 878 structures and, when found to pose a health risk, the EPA demolished and rebuilt or provided financial compensation for the structures. In total, 34 structures were addressed and contaminated soil from 18 yards was removed.

OBJECTIVE 2. *Assessment of contaminated water sources, and provision of alternative water supplies:* The EPA, NNEPA, IHS, Navajo Department of Water Resources (NDWR), and the Center for Disease Control and Prevention (CDC) exceeded the Plan goal of testing 70 rural Navajo water sources for unsafe levels of radiation or radionuclides. Of the 240 unregulated water sources tested, 29 sources exceeded the drinking water standard for uranium or radionuclides. It was beyond the scope of this effort to determine whether the source of uranium and other elements was from naturally occurring sources or due to mining and milling. With the support of Navajo Nation Chapter officials, three wells were shut down. Working together, the EPA, NNEPA, the Dine Network for Environmental Health (DiNEH) Project, CDC and the University of New Mexico conducted a comprehensive public outreach campaign including meeting with Navajo Nation Chapter officials, posting warning signs regarding health risks at water sources that exceeded the drinking water standard for uranium and radionuclides, and announcements in the Navajo Times and local radio stations.

IHS, the EPA, and Housing and Urban Development (HUD) have partnered to provide $26.7 million for 14 projects that will provide piped water to 808 homes. The Sweetwater transmission main extension also benefits 1,017 homes currently connected to piped water by improving the quality of the drinking water supply. More than $21 million of the projects funded are substantially complete which means that construction has been completed. The EPA provided $2.6 million to the NDWR to implement a water hauling program to serve up to 3,000 residents that live in remote areas.

OBJECTIVE 3. *Assessment of abandoned uranium mines, with detailed assessments of those most likely to pose environmental or health problems*: The EPA and NNEPA exceeded the Plan goal of conducting on-site screening evaluations of 200-250 mines by surveying 521 mine claims. The agencies also conducted more detailed assessments at 45 high priority mine areas, exceeding the Plan goal of conducting more detailed assessments at 35 mines.

OBJECTIVE 4. *Cleanup of the North East Church Rock (NECR) mine site and additional high- priority abandoned mine sites*: In 2011, the EPA issued a cleanup plan for the removal of approximately one million cubic yards of mine waste from NECR, three years after the target date established in the Five-Year Plan. The additional time was needed for extensive consultation on the cleanup plan with the Navajo Nation and nearby community. In the intervening time, the EPA and the responsible party conducted two large-scale interim cleanup actions to remove a total of 130,000 cubic yards of contaminated soil and addressed all known soil contamination remaining on the reservation from the adjacent NECR mine site. For the next few years, the EPA will be working closely with NRC, DOE, and the Navajo Nation to execute the cleanup plan and potential disposal of the mine waste at the UNC Church Rock Mill Site.

In addition to significant cleanup actions at the NECR Mine, the EPA and NNEPA worked together to identify mines to begin cleanup actions. These mines were selected based on radiation levels, proximity to residents, and proximity to sensitive areas such as drainage. areas that could cause migration of contamination. Between 2007 and 2012, the EPA took response actions at 9 mine sites throughout the Navajo Nation. These response actions ranged from fencing and stabilization of waste to full removal. I n 2011, mine cleanup was completed at Skyline Mine, in Monument Valley, Utah.

OBJECTIVE 5. *Cleanup of the Tuba City Highway 160 site:* The Highway 160 Site is located approximately four miles northeast of Tuba City, Arizona, on the north side of Highway 160. DOE met the Plan goal of characterizing the Highway 160 site.. In addition, DOE transferred $4.4 million to the NNEPA through a cooperative agreement to conduct remediation of the site. Navajo Nation started the project in June 2011; excavation and transportation of contaminated soil materials was completed on August 29, 2011. The material was shipped to the DOE Grand Junction Disposal Site. Backfill of the Highway160 site was completed in October 2011 and a completion report was developed. DOE performed site characterization and also provided oversight and independent verification of the remediation. At DOE's request, NRC reviewed the completion report and agreed that the site met the 40 CFR 192 cleanup standards. DOE work on remediating two structures is expected to be completed during the next five years.

OBJECTIVE 6. *Cleanup of the Tuba City Dump:* The agencies made substantial progress but did not meet the schedule established for assessment and cleanup of the Tuba City dump. Additional work and time were required due to the complex nature of site hydrology and geology. The BIA accomplished several interim actions in 2009 and 2010, including: fencing the perimeter of the dump site using funds from IHS, quarterly groundwater monitoring, wellhead protection studies to evaluate methods for protecting water supply wells, and investigation of the potential source of elevated uranium in groundwater at the dump site. All work was accomplished in consultation with the Navajo Nation and Hopi Tribe.

The Remedial Investigation/Feasibility Study (RI/FS) will provide the scientific basis for remediation of the groundwater contamination. In September of 2010 EPA and BIA signed an Administrative Order on Consent (AOC) for EPA oversight of BIA work on the RI/FS. BIA, the EPA, and the Navajo and Hopi Tribes meet regularly either by conference call or in person to review progress of the RI/FS and determine the necessary path forward. By the end of 2012, BIA had spent more than $7.6 million on the Tuba City Dump project

OBJECTIVE 7. *Remediation of groundwater contamination at three (of the four) former mill sites:* Four uranium milling sites (Shiprock, NM; Tuba City, AZ; Monument Valley, AZ; and Mexican Hat, UT) are located within the Navajo Nation's lands. DOE's Office of Legacy Management met the Plan goal of continuing maintenance of existing groundwater treatment activities at the three inactive uranium milling sites and long-term surveillance and maintenance at four inactive milling sites, all of which have been remediated by DOE Environmental Management with NRC concurrence. DOE will continue to monitor the disposal cells to ensure they remain effective in protecting human health and the environment. NRC will continue to conduct oversight activities for the general license DOE holds for the sites. DOE will continue to fund the Cooperative Agreement with the Navajo Nation for activities conducted at these sites.

OBJECTIVE 8. *Case control studies of health risks faced by individuals residing near mill sites or abandoned mine sites:* IHS and the Agency for Toxic Substances and Disease Registry (ATSDR) met the Plan goals, including diagnosing and treating known health conditions in eligible Indians, supporting a university-led Navajo Uranium Assessment and Kidney Health Project, reviewing existing databases to develop plans for improved cancer case surveillance, reviewing water contamination data for potential future health studies, and developing plans to assess the prevalence of cancer and other health conditions for populations near mine and inactive mill sites.

In 2010, IHS created the Community Uranium Exposure Journey to Healing program and has since conducted 22 community-based medical screening events. During these events 699 individuals presented themselves for examinations, and 578 of these individuals self-identified current or past non-occupational exposure to uranium.

Particular activities have included the following: training provided in 2009-10 by the CDC and ATSDR to Navajo Area IHS (NAIHS) and Navajo Nation hospital and clinic staff on uranium exposure, creation by IHS in 2010 of the community-based medical screening program Community Uranium Exposure Journey to Healing, the National Institutes of Health

(NIH) grant to the University of New Mexico for the Navajo Uranium Assessment and Kidney Health Project, and ATSDR's 2010 cooperative research agreement with the University of New Mexico for a prospective birth cohort study of environmental uranium exposure in Navajo Nation.

2. Results of Assessment and Characterization Efforts

Based on the information described above, the agencies have gained a good understanding of the scope of potential exposure to uranium contamination on the Navajo Nation and can better prioritize the work that remains to be done.

For Navajo Nation uranium mine claim screening, the EPA considered contamination greater than twice the naturally occurring levels (background levels) to be evidence of an observed hazardous release that may require further investigation under the Comprehensive Environmental Response, Compensation, and Liability Act (CERCLA; 42 U.S.C. §9601, et seq.). Of the total mine claims screened, 71 mine claims show levels of gamma radiation at less than 2 times background levels. These areas should pose little or no current threat to residents.

Of the total mine claims screened, 177 mine claims show gamma radiation levels above two times but below 10 times background levels. Long-term exposure to soils at these mines should be avoided. Residents should not build homes, corrals or other structures, and should not gather building materials from these sites. The EPA and NNEPA have initiated outreach to residents in these areas to provide warning of these conditions, and this work will continue beyond the Five-Year Plan period.

Two hundred twenty six (226) mine claims show gamma radiation levels higher than 10 times background levels. Residents should stay away from these areas. Cleanup actions have been initiated at those mines posing the highest risks, and additional actions will be necessary. Along with outreach to residents, short-term actions such as fencing, signage, or stabilization are being taken where needed.

Proximity of mines to homes is an important factor in determining risk to residents. Seventy (70) mine claims exceeding two times background levels of radiation are located within a quarter mile of a potentially inhabited structure; 36 of these mine claims have levels of radiation over 10 times background.

3. Collaboration among Federal and Other Agencies

Significant and active agency collaboration between federal, tribal, and academic partners was a key part of achieving success during the five-year period. Multiple agencies worked together on nearly every aspect of the Plan, including assessment and cleanup of contaminated structures, water supplies, mines, mill sites, the Tuba City Highway 160 site, and the Tuba City Dump. Agencies shared financial resources, including over $26.7 million provided by IHS, the EPA, and HUD for drinking water projects. Agencies have also begun innovative efforts to integrate community health screening events with mine cleanup actions to address both environmental and human health questions by community members. Finally, a Pre-Design Steering Committee comprised of staff from EPA, the New Mexico

Environment Department, the Navajo Nation, DOE and the NRC was formed in 2011 and meets quarterly to resolve technical, regulatory, and policy issues for disposal of mine waste and for enhancing the tailings cell pre-design for the UNC Church Rock Mill Site.

Beginning in 2008, four Multi-Agency Uranium Stakeholder Workshops were held on the Navajo Nation. The intent of the workshops was to bring Navajo residents together to hear about the work going on to address contamination and more importantly to hear from them about their greatest concerns. Each of these workshops were attended by community members and stakeholders and provided valuable opportunities for residents to impact the decisions about how best to address contamination on the Navajo Nation. A 5th workshop is planned for April 2013 in Gallup, New Mexico. The primary focus of this upcoming workshop will be to solicit input on future plans.

During the upcoming years, collaboration and information sharing could be further improved by conducting additional multi-agency outreach events in communities impacted by uranium mines. These events would provide increased opportunity for community members and agencies to discuss both health and environmental questions and to develop effective solutions for reducing exposure to contaminants.

4. Remaining Immediate and Longer-Term Steps to Address Uranium Contamination

Each agency has identified immediate and longer-term steps to address uranium contamination in and around the Navajo Nation. Significant areas of focus will include cleanup of additional mine sites that pose a threat to human health and the environment, assessment of additional structures and water supplies in response to new information and community concerns, development, completion and implementation of cleanup plans for the NECR mine and Tuba City dump, and completion of the prospective birth cohort study.

Challenges include the identification of suitable disposal sites for mine waste and resources for cleanup of mines lacking viable potentially responsible parties (PRPs). In May 2012, the Navajo Nation issued an executive order creating a Uranium Task Force. The Task Force is comprised of officials from four Navajo agencies and ultimately will create a uranium commission. One goal of the commission will be to explore options for safe disposal of uranium waste. The EPA will support Navajo Nation in this effort as requested. The agencies will consult with the Navajo Nation and will seek community input on work needed as future plans are developed between January 2013 and July 2013.

5. Resource Availability and Needs

The agencies largely accomplished the stated goals of the Five -Year Plan within available resources supplemented by responsible parties. In total, the agencies expended close to $110 million for work associated with the Five-Year Plan and responsible parties contributed an estimated $17 million for mine cleanup actions during this time frame.

Each agency has begun the process of identifying additional work and resources that will be needed after 2012, although significant uncertainty remains in some areas. Identification of PRPs for abandoned mines on the Navajo Nation is essential in order to provide additional resources to conduct further investigations and clean up at mines. To date, the EPA has identified and notified potentially responsible parties for 74 mine claims.

OBJECTIVE 1. ASSESSMENT AND CLEANUP OF CONTAMINATED STRUCTURES

Uranium mining or milling waste was occasionally used as sand for aggregate (foundations and stucco) and stones incorporated into the walls and floors of structures, including homes.

Structures may also be contaminated by the presence of mined or naturally-occurring radioactive materials in outside dust and soil brought into homes on shoes and clothing. Also, native soil and bedrock with elevated uranium concentrations can contribute to elevated radiation readings in and around structures. If contaminated structures are occupied, there is a risk to inhabitants from alpha radiation (e.g. radon gas) which is a potent lung carcinogen as well as gamma radiation.

In 1978 Congress passed the Uranium Mill Tailings Radiation Control Act (UMTRCA; 42 U.S.C. §7901, et seq.) to address the environmental contamination from decades of uranium ore processing at uranium milling sites mostly throughout the western United States. UMTRCA mandated remedial action for over 5,000 "vicinity properties" that contained contamination exceeding EPA Standards when contamination came from the former uranium milling site. In addition to conducting remedial action on the milling sites, DOE also remediated contaminated soils on properties surrounding and in the vicinity of the milling sites, as a part of the Vicinity Properties Program (VPP). DOE conducted investigations for contaminated properties through aerial surveys, mobile scans, and advertising. DOE's investigations were targeted based on results of mobile scans conducted by EPA and the Atomic Energy Commission in 1971 on over 1000 properties on the Navajo Reservation. Participation in the VPP was voluntary, and DOE conducted a radiological survey upon request. DOE performed detailed surveys on 87 properties, found 31 of those properties to exceed standards, and remediated those properties for a total cost of $3.5 million.

At the request of the Navajo Nation, in August 2007, the EPA Superfund Program compiled a Comprehensive Database and Atlas which identified all known abandoned uranium mines on the Navajo Nation, including information about the proximity o f t h e m i n e to water sources and structures, and the mine production history and volume estimates. This information was then used to target the structures and yards needing further evaluation. As part of the Five - Year Plan, the EPA, working with the Navajo Nation, assessed radiation levels and cleaned up, where appropriate, residential yards and homes.

Five-Year Plan Goal[1]

"In coordination with the Navajo Nation, EPA will assess up to 500 structures on their land that are likely to be contaminated with radiation at elevated levels. Where contaminated structures are found that pose a health risk to current or future occupants, EPA will take appropriate response actions using its Superfund authority. Actions may include demolition of a structure or a portion of a structure, construction of a replacement structure, or other actions to protect current and future residents from exposure to elevated levels of radiation".

Summary of Actions

The EPA and NNEPA agreed to assess homes in each mining region of the Navajo Nation, including homes within 0.25 miles of an abandoned uranium mine. Many of the home sites surveyed often had more than one structure, including homes, storage units, hogans, and other various structures. T h e EPA and NNEPA surveyed a total of 878 structures using an initial screening process. The purpose of the initial screening was to identify whether homes were contaminated or potentially contaminated with radioactive materials. After the first year of capacity building with the EPA, NNEPA Superfund Program was enabled to perform initial screens for the next four years. NNEPA surveyed 763 structures and referred 527 structures to the EPA for further analysis. The EPA's analysis involved using more sensitive equipment to determine whether contamination or potential contamination resulted in unacceptable exposure to residents. Structures then targeted for remediation contained materials from abandoned uranium mines that posed a potential health risk to current and future occupants. Results of the EPA's analysis determined the best action for remediation of the structure, and if necessary the surrounding yard. To date, the EPA has found that 43 structures and 18 residential yards pose a health risk requiring remediation.

The EPA has addressed 34 structures and 18 residential yards, with remediation ranging from demolition of an entire structure to removal of layers of soil elevated levels of uranium or other radionuclides. Cash settlements occurred in situations where there were multiple family members who could not split up a home or the home was presently unused. Temporary housing was provided by the EPA during the remediation process.

Summary of Results

The surveys and remedial actions were performed in a geographically-based phased approach. Surveys and remedial activities for Phases I, II, and III are complete; Phases I, II and III included the following areas on the Navajo Nation: Church Rock, Baca-Haystack, Red Valley, Cove, Teec Nos Pos, Cane Valley and Tuba City. Surveys for Phase IV are complete and remedial activities are underway; Phase IV survey areas included Tuba City, Cameron, and Bodaway/Gap. Surveys for Phase V occurred during Fall/Winter 2012; Phase V survey areas include Indian Wells, Blue gap, Tselani, Chinle, Chilchinbeto, Black Mesa, Oljato, and Dennehotso.

The completed survey results are summarized below in Table 1.

Table 1. Navajo Nation Contaminated Structure Assessments

Actions	Total
Total Structures Surveyed (by the EPA and NNEPA)	878
Structures Confirmed Contaminated to Date	43
Structures and/or Residential Yards Remediated by the EPA	52
Structures and/or Residential Yards Pending Remediation by the EPA	9

Structure Near Mine Site.

Structure Demolition.

Rebuilt Structure.

OBJECTIVE 2. ASSESSMENT OF CONTAMINATED WATER SOURCES AND PROVISION OF ALTERNATIVE WATER SUPPLIES

Nearly 30% of the Navajo population obtains drinking water from unregulated sources, such as livestock wells, springs, private wells, or watering points. Navajo Nation policy prohibits the use of these unregulated sources for human consumption, but the practice continues due to lack of suitable water sources in the more remote areas. Many unregulated sources contain naturally occurring levels of uranium, arsenic, and other elements. The use of unregulated water sources represents the greatest public health risk associated with drinking water for the Navajo Nation.

Regulated public water supplies are currently monitored by NNEPA, and levels of uranium and other radionuclides are within safe drinking water standards.

Five-Year Plan Goal

"In spring 2008, the EPA will test up to 70 rural Navajo water sources such as livestock wells, windmills, artesian wells and springs to determine if they contain safe levels of radiation or radionuclides.

T h e EPA will work with the Navajo Nation, the IHS, and other agencies to provide the results to the affected communities and develop plans to ensure access to clean and safe water for drinking. The EPA will use Superfund authority to take appropriate action at water sources contaminated by mine waste and found to be endangering human health".

Summary of Actions

The EPA, NNEPA, IHS, NDWR, and the CDC tested 240 unregulated water sources for elevated uranium or radionuclides. The exact number of unregulated sources is unknown, but estimated to be in the low thousands. Water sources were prioritized for sampling based on proximity to abandoned uranium mines, existing information indicating potential contamination, or other known uses. Regulated water sources were not sampled because these sources are routinely monitored in accordance with the requirements of the Safe Drinking Water Act.

Summary of Results

Of the 240 unregulated water sources tested, 29 sources exceeded drinking water standards for uranium or radionuclides. Durable signs were posted at these contaminated water sources warning individuals of the health risks associated with hauling water with elevated radionuclides and other minerals. With the support of Navajo Nation Chapter officials, three contaminated wells were shut down. It was beyond the scope of this effort to determine whether the source of uranium and other radionuclides was from naturally occurring sources or due to mining and milling. Working together, NNEPA, the EPA, the

DiNEH Project, CDC and the University of New Mexico conducted a comprehensive public outreach campaign regarding these wells. Outreach included meeting with officials of affected Navajo Nation chapters, the posting of yellow warning signs, and the publication of outreach flyers in the Navajo Times.

In addition, NNEPA, the University of New Mexico, the DiNEH Project and the CDC aired radio announcements warning residents about the dangers of using livestock water sources. NNEPA and the EPA issued a joint press release informing residents that only regulated water sources are safe to drink because they are regularly sampled by the Navajo Tribal Utility Authority and they meet the Safe Drinking Water Act standards.

The NAIHS Office of Environmental Health, the Navajo Nation and the EPA collaborated to identify homes within a 10 mile radius of abandoned mines or stock wells known to have high levels of uranium. From this list, 13 water/sewer infrastructure projects were targeted for funding from 2009-2012 in Arizona and New Mexico plus a Black Falls, Arizona, pilot project which was completed in 2010. The pilot project funded by EPA provided NAIHS designed solar powered cistern pump water delivery to six family homes instead of the more conventional designs of gravity fed and electrical powered cistern pumps. To date there have not been any reports of failures for the solar powered systems.

IHS, the EPA, and HUD have partnered to provide $26.7 million for 14 projects that will provide piped water to 808 homes. The Sweetwater transmission main extension also benefits 1,017 homes currently connected to piped water by improving the quality of their drinking water supply. The EPA provided $2.6 million to the NDWR to implement a water hauling program to serve up to 3,000 residents that live in remote areas. NDWR delivered the first loads of water in April 2011, and is working with 328 residents in 40 Navajo Nation chapters.

Four trucks are currently delivering water. The EPA worked with the Navajo Tribal Utility Authority to fund the construction of new regulated watering points in several Navajo Nation chapters, including Baca, Red Mesa and Coalmine Mesa Chapters. NNEPA also developed safe water hauling guidelines which are posted on their website: www.navajopublicwater.org. NNEPA sponsors workshops for the Community Health Representatives (CHRs) who are within the Navajo Nation Division of Health (NNDOH). These workshops help the CHRs recognize potential health risks and identify for medical referral any individuals who exhibit health problems that may be caused by environmental sources, including drinking water from unregulated sources or living near an abandoned uranium mine site. These workshops are well received and the CHRs, who are often the first to provide outreach when they visit people at their homes, can now provide better assistance.

The water sources that exceeded drinking water standards and improvement projects are detailed in Tables 2 and 3.

Limitations

- Navajo Nation Chapters are reluctant to close contaminated wells because funds to replace these wells are very limited. Some wells were disabled only to be reconnected by residents for livestock use.

Table 2. Navajo Nation Unregulated Water Sources
Twenty-nine of the 240 sources sampled contained constituents above drinking water standards, as summarized below

Water Source	Navajo Nation Chapter	Uranium (µg/L) MCL30 µg/L	Gross Alpha (pCi/L) No MCL	Radium226+228 (p/Ci) MCL5 p/Ci	Arsenic (µg/L) MCL10 µg/L	Lead (µg/L)15 µg/L action level	Selenium (µg/L) MCL50 µg/L
Badger Springs	Cameron	39			31		
3A-155TohatchiSpring	Cameron	120			40		
3A-PHS-32Paddock Well	Cameron	55			25		
16-4-10Lime Ridge	Church Rock	260	108	9.6			
16K-336	Church Rock			5.78	11		
Becenti TrailSpring	Church Rock	110					
17M-99	Cornfields	100					
12-7-12Ellison Well	Cove	61					
8A-180	Dennehotso	31					
8A-179	Dennehotso	46					
17M-66	Ganado	31			18		
17T-559	Greasewood	78					
17-8Snake Well	Greasewood	32			20		
16B-38Paddy Martinez	Haystack	54					
16T-521Platero	Haystack	63	20.6		55		
17H-146	Indian Wells	69			51		
Pigeon Springs	Iyanbito	88.5					
5M-74Box Springs	Leupp	35					
10T-241A	Many Farms	44					
Monument Pass	Oljato	39	16.37		11		
8K-433	Oljato	130		9.57	11		
17T-519	Steamboat			8.19			
17T-545	Steamboat	32.85			26.8		
9Y-12	Red Mesa / Mexican Water	700				68	140
9Y-32	Red Mesa	51			21		
16T-519 Largo Corral	Smith Lake	34			32		
9T-550	Sweetwater	31	18.93	6.21			
9T-586	Sweetwater		22.43				
10R-51B	Tselani/Cottonwood	31					

Note that the water sources cited were not sampled from Public Water Supply Systems. The MCLs were used for comparison purposes only. The results are not definitive with respect to attribution from mining versus naturally occurring sources.

Table 3. Navajo Nation Water Infrastructure Projects

Project	Project Completion Status	Homes Served
Monument Valley waterline extension	6/12 substantially complete	128
Black Falls waterline extension and watering point	9/10 complete	18
Dennehotso new water system	6/12 substantially complete	107
Sweetwater transmission main extension	6/12 substantially complete	1,109
Pinedale scattered sites	3/12 complete	5
Church Rock scattered sites	1/12 complete	4
Haystack scattered sites	6/12 construction started	6
Mariano Lake North waterline extension	6/12 complete	62
Leupp / Grand Falls Test Well	Planning/design underway	Unknown
Red Mesa Water Hauling point	6/12 complete	Unknown
Mexican Water Walker Creek Extension	Planning/design underway	38
Thoreau scattered extension	Planning/design underway	9
Peretti Canyon	Planning/design underway	11
Water Hauling	Water hauling began 4/11. 328 homes in 40 chapters are being served.	328
Total		**1825 homes served (808 homes served with piped water for the first time**

Contaminated Water Sources.

Water Hauling.

New Infrastructure Construction.

Objective 3. Assessment of Abandoned Uranium Mines, with Detailed Assessments of Those Most Likely to Pose Environmental or Health Problems

Prior to 2007 the EPA identified approximately 1,100 mine features (portals, prospects, rim strips, pits, vertical shafts or waste piles) at 520 discrete abandoned mine claims throughout the Navajo Nation. Based on this information, t h e EPA compiled a Comprehensive Database and Atlas which provides detailed maps, photos and other relevant data about these mine claims. The Atlas is available at http:/www.epa.gov/region9/superfund/navajo-nation.

The Navajo Abandoned Mine Lands Program (AML) used Surface Mining Control Reclamation Act (SMCRA; 30 U.S.C. §1234, et seq.) funds to address an estimated 90% of the surface hazards posed by these abandoned uranium mines (AUMs). Some of the mines continue to release contaminants into the environment and will require further cleanup or future O&M. The AML work included reclamation of 913 abandoned uranium mine features identified. This information is available at http://www.aml.navajo-nsn.gov/AML_Files/AMLAccomplishmentsPage.html.

Five-Year Plan Goal

"EPA and the Navajo Nation will work together to conduct a tiered assessment of abandoned mines, with more detailed assessments of those most likely to pose environmental or health problems. In 2007, EPA completed a geographic information system inventory and risk-ranking of the 520 known mines. In 2008, EPA will use this list and consult with the Navajo Nation to identify the highest-priority 200-250 mines for further evaluation. EPA expects to conduct on- site screening evaluations of these mine sites, at the rate of up to 50 per year through 2012. EPA also expects to conduct more detailed assessments of up to 35 of the highest priority mines by 2012, at the rate of approximately seven sites each year."

Mine Assessments

Summary of Actions

While the plan called for 250 mines to be evaluated, the EPA determined that all mine claims should be surveyed. Three additional mine claims were identified and a total of 521 mine claims were assessed from October 2008 until November 2011. The EPA conducted on-the-ground gamma radiation screens at all accessible identified mine claims. A total of 474 mine claims were scanned for gamma radiation. Forty-seven of the mine claims were not scanned for gamma radiation due to access issues. Note that often several mines are located within a mine claim. A mine site screen report was completed for each mine claim. The report includes a summary of the field observations, a review of the database information, site photographs, and figures showing the gamma radiation measurements throughout each site. The screening reports along with data from the Atlas are being used to help determine which sites have the greatest potential to impact surrounding communities. The reports were used to determine which sites should be further evaluated for listing on the National Priorities List or if they should be referred to the EPA's emergency response program. The screening reports are available at http:/www.epa.gov/region9/superfund/navajo-nation/.

Following the field survey, the data was integrated into the Geographic Information System to be used for mapping and evaluation. Specific mine features, such as portals, rim strips, pits, and waste rock piles, were recorded into the field observation notes as well as collected as GPS waypoints. In addition, any observed structures, potential drinking water sources, and surface water drainages were identified and recorded. Site photographs were also taken at each mine. As part of scanning the mine claim, the EPA collected off-site gamma readings for each mine in order to evaluate the results against naturally occurring gamma radiation levels in the vicinity.

Extensive coordination was conducted with Navajo tribal officials, Navajo community members, state agencies, and private land owners. Prior to the site visits, meetings were held with local officials and residents at various Navajo Nation Chapter Houses to discuss site access issues. Site access meetings were held with State officials from the New Mexico Energy, Minerals, and Natural Resources Department, Mining and Minerals Division and the New Mexico Environment Department. Local property owners and ranchers were also consulted with in order to gain access to several of the sites.

Based on results of the screening reports, more detailed assessments were completed at 45 mines. The assessments evaluated exposure factors in order to determine whether any of the mines could be eligible for potential listing on the National Priorities List or if they should be referred to the EPA's emergency response program for an immediate action. In July 2008, Preliminary Assessments were completed at several of the mines which had been previously investigated and entered into the EPA's Comprehensive Environmental Response, Compensation, and Liability Information System database and required further evaluation using the Hazard Ranking System. A Site Inspection was completed for the King Tutt Mesa Aggregated Site, which is a mining complex of more than 30 mines. The EPA is currently working on two additional Preliminary Assessments; one for the Section 9 Lease located near Cameron, AZ and an updated assessment for the Mesa I Mines and Cove Wash areas based on new information provided from the screening data. The detailed assessments are summarized below in Table 4. Two of the mine claim areas are currently undergoing cleanup actions as described in the next section.

Table 4. Detailed Mine Assessments

Mine Claim	Region	Assessment Action	Current Status
Billy The Kid Mine	Eastern	Preliminary Assessment	Assessment completed in 2008; not eligible for listing
Cove Transfer Station	Northern	Preliminary Assessment	Referred to Response Program; cleanup in progress
Mariano Lake	Eastern	Preliminary Assessment	Referred to Response Program; cleanup in progress
Moonlight Mine	NorthCentral	Preliminary Assessment	Assessment completed in 2008; not eligible for listing
King Tutt Mesa Mines	Northern	Reassessment	High priority mines to be re-evaluated in 2013
Section 9 Lease	Western	Preliminary Assessment	Assessment in progress
Mesa Mines/ CoveWash	Northern	Preliminary Assessment	Updated assessment in progress

Summary of Results

Based on the information described above, t h e EPA has a good understanding of the scope of potential exposure issues related to uranium mining on the Navajo Nation. Health effects as a result of chronic uranium exposure via contaminated soil include lung cancer, bone cancer and impaired kidney functions.

Of the total mine claims screened, 71 mine claims show levels of gamma radiation at less than 2 times background levels. Areas with levels that are at or below 2 times background levels should pose little or no current threat to residents.

For mine claim screening, the EPA generally considered contamination (including radiation levels) greater than twice the naturally occurring levels (background levels) to be evidence of an observed hazardous release that may require further investigation under CERCLA.

Of the total mine claims screened, 403 mine claims show gamma radiation levels above 2 times background levels. Long-term exposure to soils at these mines should be avoided. Residents should not build homes, corrals or other structures, and should not gather building materials from these sites.

Of the 403 mine claims showing gamma radiation levels above 2 times background levels, 226 mine claims show gamma radiation levels higher than 10 times background levels. Residents should stay away from these areas. The EPA and Navajo EPA have begun working closely with Navajo Nation Chapters to warn them of areas with elevated levels and are working together with residents to determine whether warning signs and/or fencing would be appropriate to keep people away from these areas.

Of the total mine claims screened, 70 mine claims show levels of radiation greater than 2 times background and are located within .25 mile of a potentially inhabited structure; 36 of which have levels of radiation over 10 times background.

The EPA and NNEPA have been using the screening information described above along with results from the Atlas to determine where to allocate resources first; beginning with mines that are near structures. EPA is conducting CERCLA response actions at 9 of the 36 mine claims that have levels of radiation more than 10 times background levels and are within 0.25 miles of structures. As future plans are developed, information from the mine claim screening will be used to prioritize work.

Limitations

Forty-seven mines were not accessible, and therefore gamma radiation measurements and field observations were not collected. However reports for these mines were written and provide as much information as could be gathered from the Atlas and visual observation.

DOE will continue to work with the EPA to provide information on the results of historical vicinity property surveys and abandoned mine properties to determine the extent of other contaminated properties on Navajo Nation lands.

AUM Screening.

AUM Screening.

OBJECTIVE 4. CLEANUP OF NECR MINE SITE AND ADDITIONAL HIGH-PRIORITY ABANDONED MINE SITES

Cleanup of NECR Mine

Located north of Gallup, New Mexico and on Navajo Nation land, the NECR mine is the highest priority mine site cleanup for the Navajo Nation and t h e EPA. NECR was formerly operated by the United Nuclear Corporation/General Electric (UNC/GE).

Five-Year Plan Goal

"The Northeast Church Rock Mine located near Gallup, New Mexico, is the highest priority cleanup on EPA's AUM ranking list. EPA will determine the soil remedy in 2008 and plans to require United Nuclear Corporation, now a subsidiary of GE, to perform a comprehensive Superfund removal action for cleanup of soils on the balance of the site. Cleanup is expected to take up to three years. Costs are expected to be provided by the responsible party."

Summary of Actions

The EPA issued a cleanup plan for the removal of approximately one million cubic yards of mine waste from NECR, and continued to conduct removal actions to address soil contamination at the site and throughout the surrounding community.

The cleanup plan uses one of the most stringent uranium mine cleanup standards in the country, and details the intent of sending waste containing high levels of radium or uranium off-site for reprocessing or approved disposal, adding a cap and liner system for the mine waste to potentially be disposed of at the UNC Church Rock Mill Site to ensure that the mine waste does not affect people and the environment. The NECR mine is located on Navajo Nation Trust land and the UNC Church Rock Mill Site is located on fee land held by UNC and is regulated and licensed by the NRC. UNC is now an indirect subsidiary of General

Electric Corporation (GE). The plan provides for unlimited surface use of the mine site after cleanup, voluntary housing options during the cleanup for community members living near the mine, and job training and employment during the cleanup.

EPA, DOE, NRC, State of New Mexico, and Navajo Nation have been working closely on the cleanup plan for the NECR Mine Site. The mine site is being remediated by UNC/GE under the CERCLA. Implementation of the cleanup plan is contingent upon the EPA's Record of Decision for the UNC Church Rock Mill Site anticipated by March 2013 and license amendment approval from the NRC anticipated in 2016. The cleanup action calls for the mine waste contaminated material to be placed in a new disposal cell to be built on top of the existing Church Rock uranium mill tailings disposal cell. The design of the disposal cell and placement of the mine waste at the UNC Church Rock Mill Site would require NRC approval of a license amendment request from the site owner. DOE has reviewed the EPA's proposed plan and has provided detailed comments on the approach for mine waste disposal. If the mine waste is co-located with the mill tailings at the UNC Church Rock Mill Site, after GE/UNC completes remediation and NRC approves termination of the current licensee's radioactive materials license, DOE likely will conduct long-term surveillance and maintenance (LTS&M) for the site under UMTRCA. DOE will continue to work with the Navajo Nation regarding the reclamation and transfer of this site for long term management to ensure protection of human health and the environment.

During the five year planning period, the EPA direct and indirect oversight costs have exceeded $4 million which include labor, contractor support, and the cost of temporary housing for residents during removal actions. GE has reimbursed the EPA for past costs. In addition, the cost to GE has exceeded $6 million for interim removal actions as well as progress on the long-term action. GE/UNC entered into an agreement with the United States (DOE/DOI) to reimburse GE/UNC for approximately one-third of NECR-related costs.

Summary of Results

In 2009, the EPA cleaned up more than 100,000 cubic yards of contaminated soil on the Navajo Nation around the NECR Mine. The waste was consolidated on the existing mine waste pile which was also re-graded and covered to prevent further migration of waste before the final cleanup of the mine site. In 2011, an additional area of contaminated soil outside the NECR fence line on the reservation was discovered. A fence was installed around the area immediately to control access and exposure. Beginning in October 2012, GE under the EPA's oversight, removed approximately 30,000 cubic yards of contaminated soil. This soil was moved to the existing waste pile at NECR and covered with clean soil. During the same time period, 15,000 cubic yards of contaminated soil was removed by Rio Algom (the responsible party for Quivira mine) from a drainage area adjacent to NECR and placed on the waste pile currently located at Quivira mine. Seventy residents were temporarily relocated during these actions. All interim cleanup actions were completed, and residents returned to their homes on November 17, 2012.

Factsheets, action memos (cleanup decisions), technical documents, enforcement documents, public comment documents, and administrative records can be accessed at www.epa.gov/region09/NECR.

Limitations

- Implementation of the cleanup plan is contingent upon the EPA's Record of Decision for the UNC Church Rock Mill Site anticipated by March 2013, and license amendment approval from the NRC, anticipated in 2016.

Cleanup of remaining contaminated soil outside the NECR fence line, completed November 2012.

Cleanup of NECR Mine.

Cleanup of Additional High Priority Mines

Summary of Actions

In addition to significant cleanup actions at the highest priority mine site, NECR Mine, the EPA and NNEPA worked together to identify mines at which to begin physical response actions. These mines were selected based on radiation levels, proximity to residents, and proximity to sensitive areas such as drainage areas that could cause migration of contamination.

Summary of Results

Between 2007 and 2012, the EPA began response actions at 9 mine areas throughout the Navajo Nation. In 2011, the first full mine cleanup was completed at Skyline Mine, in the North Central Region. Cleanup of Skyline mine included moving 25,000 cubic yards of radioactive mine waste to a repository near the abandoned Skyline Mine. The repository, a shallow pit, was sealed with high-density polyethylene (HDPE), then covered with local soil and rock. In addition, cleanups are underway at mines in the Eastern and Northern Regions of Navajo Nation.

The status of response actions at mines is summarized below in Table 5.

Table 5. EPA Response Actions

Mine Site	Region	Action	Status
NECR mine	Eastern	Cleanup	Third interim action occurred Oct/Nov 2012. Cleanup to occur 2016-2020. www.epa.gov/region9/necr
Mariano Lake Mine	Eastern	Investigation/ Cleanup	Urgent actions have been taken. Investigation ongoing. Evaluation of cleanup options to begin in 2013.
Quivira Mine	Eastern	Investigation/ Cleanup	Second interim action occurred Oct/Nov2012. Cleanup to occur 2016-2020. www.epaosc.org/Quivira_Mine
Skyline Mine	North Central	Cleanup	Clean up completed October 2011; total cost of $8 million. www.epaosc.org/SkylineAUM
Cove Transfer Stations	Northern	Cleanup	Investigation complete. Urgent actions occurred October 2012. www.epaosc.org/CoveTransferStations
Sections 32 and33	Eastern	Cleanup	Investigation complete. Urgent actions occurred Oct/Nov 2012. www.epaosc.org/Section32_33AUM
Ruby Mines1-4	Eastern	Negotiations with PRP	Starting negotiations with PRP. Urgent actions occurred Oct 2012.
20 Mine Claims in Cameron Area	Western	Negotiations with PRP	Starting negotiations with PRP. Site visits conducted Sept/Nov 2012.

Cleanup of Skyline Mine.

Cleanup at Section 32/33.

Repository Cell at Skyline Mine

Cleanup at Cove Transfer Station.

Repository Cell at Section 32/33.

Identification of Potentially Responsible Parties and Enforcement

Identification of Potentially Responsible Parties (PRP) for abandoned mines on the Navajo Nation is essential in order to provide additional resources to conduct further investigations and clean up at mines.

Five-Year Plan Goal

"High-priority mines will be candidates for Superfund enforcement. The EPA will continue to pursue an enforcement-first policy, and will conduct a general search for Potentially Responsible Parties at all of the mines. The EPA plans to issue information request letters to several major viable corporations (potentially responsible parties) who owned or operated multiple mines by 2008, and the EPA will continue to assist Navajo Department of Justice with Navajo-lead enforcement actions".

Summary of Actions

In 2008 the EPA sent CERCLA 104(e) letters requesting information about potential liability to 10 companies that had been previously identified as having mined uranium on the Navajo Nation. The EPA used the information provided to identify PRPs for mines posing the highest risks. Prior to initiating extensive research, the EPA and NNEPA worked together to identify mines that both agencies agreed pose the greatest risk to human health and the environment.

Summary of Results

To date, the following mines are being addressed by the EPA and NNEPA and are being paid for by responsible parties who have signed cleanup orders with the EPA:

- NECR Mine;
    - Responsible Parties: United Nuclear Corp, General Electric
- Mariano Lake Mine;
    - Responsible Party: Chevron USA, Inc.
- Quivira Mine;
    - Responsible Party: Rio Algom Mining LLC. Note that Rio Algom purchased and operated Quivira mine from the predecessor to Tronox Corporation (Kerr McGee Corp). The EPA has signed an agreement with Rio Algom to investigate and cleanup contamination associated with Quivira mine.

The EPA is paying for the cleanup of Sections 32 and 33 and the Cove Transfer Stations with funds received via the federal bankruptcy of Tronox Inc. as a successor to Kerr-McGee and will prioritize the use of any remaining bankruptcy funds at 49 other Kerr-McGee mine claim sites.

In addition to the above referenced parties, the EPA has identified and sent liability notices to two companies regarding their liability at 24 mines. The EPA is in negotiations with Freeport McMoRan Company to investigate and clean up the Ruby Mines 1, 2, 3, and 4, which are located in the Eastern Agency. The EPA is also in negotiations with Kinder Morgan, formerly El Paso Natural Gas, to investigate and cleanup 20 mine claims located near Cameron, AZ.

The PRPs that have been notified of their liability are summarized in Table 6.

Table 6. Identified PRPs

Potentially Responsible Party	No. Mine Claims
Tronox Incorporated	49
KinderMorgan, Inc. (El Paso Natural Gas Company)	20
Western Nuclear, Inc. / Freeport-McMoRan Copper and Gold, Inc.	2
United Nuclear Corporation /General Electric	1
Chevron U.S.A. Inc.	1
Rio Algom Mining LLC	1
Total Number of Mine Claims with Identified PRP	74

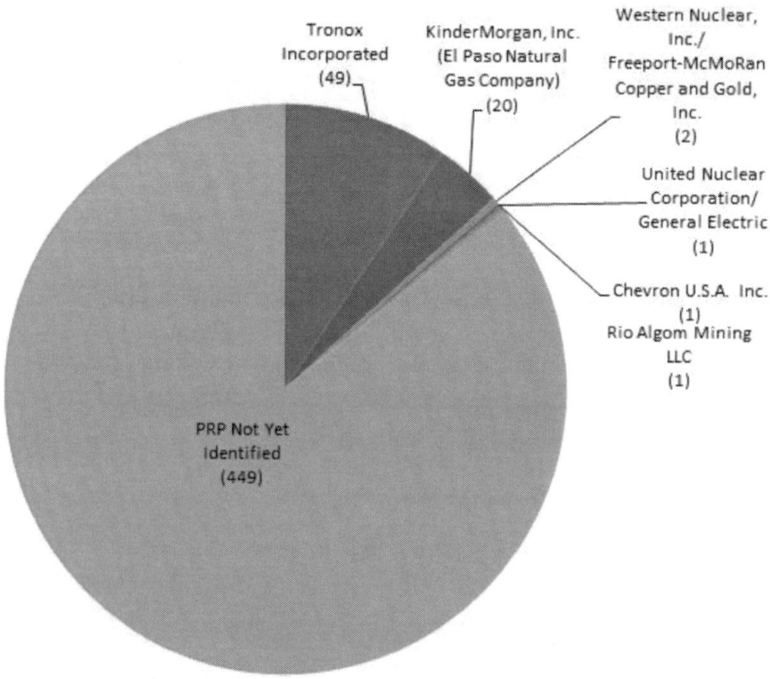

OBJECTIVE 5. CLEANUP OF THE TUBA CITY HIGHWAY 160 SITE

The Highway 160 Site is located approximately four miles northeast of Tuba City, Arizona, on the north side of Highway 160. DOE maintains a disposal cell and operates a groundwater treatment system on the south side of Highway 160 (the Tuba City Disposal Site). In the 1980s and 1990s, DOE conducted characterization and remediation of the former uranium mill site (south of Highway 160) under UMTRCA. Previous surveys conducted by DOE on the north side of Highway 160 indicated no contamination in this area. The 16-acre property is vacant land, but there are 14 residential structures within one mile of the site.

Five Year Plan Goal

"Characterize the Highway 160 Site: DOE will work with BIA, USGS, EPA, the Navajo Nation, and El Paso Natural Gas (EPNG), as appropriate, to understand the best approach for characterization and any required cleanup based on characterization of the site.

There are currently three authorized courses of action related to additional characterization and remediation of the Highway 160 site: (1) EPA cleanup of the site as a Superfund removal action; (2) continued action by EPNG; (3) some combination of the two approaches.

If EPA determines that an Imminent and Substantial Endangerment exists at this site, EPA can use Superfund enforcement authority to require the potentially responsible parties to conduct a response action. DOE will cooperate with that effort, as requested by the parties as appropriate. If El Paso Natural Gas works to affect a cleanup of this site, DOE will work with them to consider disposal of the contaminated materials at the DOE facility near Grand Junction.

DOE has provided support to the preparation of a cost estimate for the remaining cleanup work based on the EPNG characterization report. The cost estimate would require additional characterization data that provides information on the chemical species and the depth of the contamination. Based on existing site knowledge, DOE estimates it would take several years to complete the work at the Highway 160 site".

Summary of Actions

The Navajo Nation through a cooperative agreement with DOE conducted the remediation of the Highway 160 site. DOE performed the characterization and also provided oversight and independent verification of the remediation.

Because the characterization concluded the radioactive waste originated from the former mill site, it could be classified as residual radioactive material (RRM) and therefore could be disposed of at the Grand Junction Colorado Disposal Cell (GJDS) which is an UMTRCA disposal site. Congress authorized DOE to keep the GJDS open to receive RRM from properties discovered after DOE's authority to remediate RRM expired in 1998. DOE only opens the cell as needed to accommodate cleanups such as the Highway 160 site. DOE worked with the Navajo Nation to keep the cell open during the period of site remediation and RRM shipment.

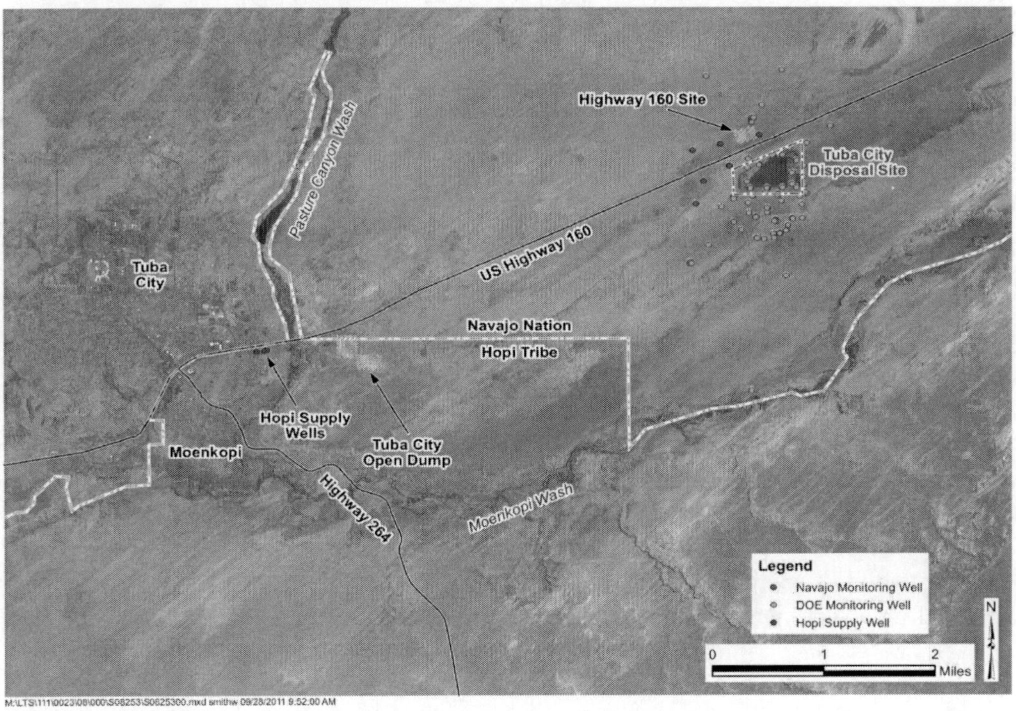

Highway 160 Site.

DOE also worked with the Navajo Nation to install wells to reevaluate the extent of groundwater contamination in the area of the Highway 160 Site and the Tuba City UMTRCA Disposal Cell; no contamination was found. Data from these wells also demonstrates that contamination is not migrating from the Tuba City Disposal Cell to the Tuba City Open Dump. Some of these wells were transferred into DOE's long-term surveillance network and continue to be sampled on a regular basis.

The entire Highway 160 project took less than one year from initial field characterization to completing remediation of the soil contamination. At DOE's request, the NRC reviewed the project completion report documenting the site cleanup and agreed that the site met the 40 CFR 192 cleanup standards. DOE is working with EPA and the Navajo Nation to use remaining funding from the project to remediate two structures that will be completed during the next five year plan.

Objective 6. Cleanup of the Tuba City Dump

The Tuba City Dump (TCD) was used for more than 50 years as an open, uncontrolled dump receiving solid waste from local Navajo and Hopi communities. The BIA serviced the TCD by periodically covering trash and burying waste with available soil. Information on the type of waste disposed of at the TCD is limited because the site was unmanned and unsecured. The TCD occupies land on both the Hopi Reservation and the Navajo Nation Reservation. The BIA, along with the EPA, and the IHS, are working collaboratively with the Hopi Tribe and Navajo Nation to address environmental concerns at the site. The contaminants of concern (COC) include persistent elevated levels of uranium, vanadium, radium 226/228, gross alpha and beta activity, metals, selenium, chlorides, nitrates, sulfates and arsenic in the shallow groundwater sources at the site. Deeper groundwater and nearby drinking water supply wells have not exhibited elevated levels of these constituents. The TCD was closed by BIA in August 1997 thereby prohibiting any further use of the site as a dump. It had been the only dump site available to the Navajo and Hopi communities of Tuba City and the Upper and Lower Hopi Villages of Moenkopi prior to that time.

Five Year Plan Goal

"The five federal agencies will work together with the Navajo and Hopi tribes in early 2008 to assess whether interim actions are needed prior to selection of a final remedy for the site. By August 2008, the Bureau of Indian Affairs plans to complete an assessment of the need and feasibility of conducting an interim measure to prevent contamination of nearby Hopi water supplies. If an imminent threat to water supplies is identified, the agencies will determine the most appropriate authorities to achieve an interim remedy. These authorities might include a Superfund response or enforcement action."

Summary of Actions

Since 1999, the BIA has been conducting assessment activities of uranium and other COCs in groundwater, springs, contaminant migration pathways, sources, and receptors for the purpose of formulating a final closure plan for the TCD. The Hopi Tribe has concerns because the springs in the area are used by the Tribe for religious and ceremonial purposes. Groundwater monitoring data suggests that COCs in groundwater may extend to the west and southwest of the site. Groundwater sampling and analyses in the TCD as well as up-gradient and down-gradient indicate shallow groundwater COCs may not be limited to the TCD site. Both tribes have consistently requested "clean closure" (removal of all waste).

BIA and the EPA have worked with the Hopi Tribe and the Navajo Nation for the last five years to resolve the problems at the TCD. In 2008, the BIA performed an interim action study prior to the CERCLA RI/FS process, to determine the movement of COCs in both deep and shallow groundwater to assess any imminent threat to drinking water supply wells, irrigation water, springs and seeps. As a result no imminent threat was found to the drinking water supply wells. However, other actions were recommended in the Interim Action Report (IAR) to safeguard human health and the environment in and around the dump. The IAR recommendations included:

1. Fencing the perimeter of the old cell to prevent direct contact with exposed waste;
2. Quarterly Monitoring of deep and shallow groundwater and spring water to confirm the findings and conclusions of the IAR;
3. Wellhead protection studies to evaluate methods for protecting water supply wells from shallow ground water intrusion;
4. Source investigations for uranium and other COCs around a "hot spot" known as monitoring well 07 (MW-07) which exhibited elevated levels of contamination.

In response to the report, in 2009 the BIA proceeded to implement all IAR recommended interim action measures. The landfill was fenced off to prevent the possibility of exposure of humans to possible COCs in the soils. NAIHS contributed $80,000 to the BIA to allow the BIA to complete the fencing. BIA funded an investigation by EPA to locate the source of uranium and other COCs in or around MW-07. Two sentinel wells were installed between the dump and the Hopi drinking water source to provide early detection in the event that the COCs started migrating toward the water supply wells. Quarterly groundwater monitoring and the wellhead protection study were contracted with the Hopi Tribe to aid in the evaluation of measures to be taken to protect the groundwater.

The EPA performed a limited waste characterization and "Hot Spot" investigation around MW-07 where uranium levels were highest. The operation was completed in five phases.

1. non-intrusive surface and down hole gamma spectroscopy surveys
2. direct push borings, core samples and analysis
3. five test pits excavated at locations to attempt to locate a uranium source
4. 8 hollow stem auger boring sites identified during phase 1 through 3 to indicate elevated uranium and for ground water flow direction and volume information
5. groundwater sampling, aquifer testing and slug testing to evaluate groundwater flow direction and water quality data

The investigation resulted in shallow groundwater COCs detected above applicable maximum contaminant levels within a fifteen foot area in the vicinity of well MW-07. Another monitoring well was created with even higher uranium readings adjacent to MW-07. The investigation did not determine the source of elevated uranium in groundwater. Additional evaluation will be conducted during the ongoing Remedial Investigation. This report is expected to be completed in 2013.

In 2010 the RI/FS Work Plan and all accompanying work plan documents such as the Sampling and Analysis Plan, Quality Assurance Project Plan and the Health and Safety Plan were written resulting in an approved RI/FS workplan. The core samples taken by the EPA during the Hot Spot investigation were stored by USGS and analyzed in 2010. The results of the analysis were inconclusive and failed to confirm the source of elevated levels of uranium contamination, although the isotopic signature of the MW-07 investigation was not consistent with the isotopic signature of mill waste or mine tailings. EPA and BIA entered into an Administrative Order on Consent (AOC) for EPA oversight of BIA work on the RI/FS.

Work on the ground water monitoring and supply well testing continued in 2011 and the RI/FS workplan was revised. Field studies such as the vegetation inventories began in 2011. Groundwater monitoring continued along with drinking water well testing.

In 2012, radiological and soil vapor surveys began. The final RI/FS work plan field investigations and data collection began along with soil borings and the hydro geologic investigations. The soil boring and hydrogeology investigations and analysis were scheduled for completion by February of 2013. However, the Continuous Multi-channel Tubing wells were delayed resulting in a delay in the data analysis phase. The RI/FS report is still on track to be completed by the end of calendar year 2013. Reviews of the draft report will be available to stakeholders prior to the release of the final report.

Future Plans

The agencies will continue to work together following directions outlined in the AOC agreement. The BIA will continue to consult with the Hopi Tribe and Navajo Nation in its decisions. The timelines are set as indicated below. All time frames are subject to adjustments dictated by the RI/FS and culminating in the ROD.

Task 1: The BIA is scheduled to submit the remedial investigation report by December of 2013 with the feasibility study report submitted in January of 2014. This will include soil sampling and analysis, data validation and continued groundwater sampling.

Task 2: EPA is slated to issue the ROD for the TCD in 2014. The ROD is the instrument EPA will use to determine the course of action for the cleanup/remediation process at the TCD. Stakeholder meetings on the ROD will be held prior to a final decision.

Task 3: The Remedial Design/Remedial Action Plan should be completed in 2015 and Closure should begin in 2015 or 2016 depending on the remedial actions decided upon. Closure timelines for the remediation will also be dependent upon the type of work required under the ROD.

Task 4: The length of time for the groundwater remediation, as well as cleanup and final closure activities, cannot be reasonably estimated at this time and will depend upon the remedial actions required in the ROD. Groundwater monitoring is planned to continue throughout the completion of the closure with the duration of monitoring also determined by the ROD.

OBJECTIVE 7. REMEDIATION OF GROUNDWATER CONTAMINATION AT THREE FORMER MILLING SITES

DOE's Office of Legacy Management is responsible for long-term care of 26 UMTRCA sites in ten states. Four former uranium milling sites (Shiprock, NM; Tuba City, AZ; Monument Valley, AZ; and Mexican Hat, UT) are located within Navajo Nation lands (see Figure 1). These sites are under the general license issued by the NRC pursuant 10 CFR 40.27.

DOE met the Five-Year Plan goal of continuing maintenance of existing groundwater treatment activities at the three inactive uranium milling sites and long-term surveillance and maintenance at three disposal cells located on the former milling sites, all of which have been remediated by DOE. DOE will continue to monitor the disposal cells and the NRC will continue its oversight of DOE to ensure the disposal cells remain protective of human health and the environment. DOE's LTS&M and the NRC's oversight responsibilities will continue for the foreseeable future.

In 1983, DOE and the Navajo Nation entered into a cooperative agreement to facilitate the surface remedial action of the Shiprock, Monument Valley, Tuba City, and Mexican Hat sites. Funds provided by DOE to the Navajo Nation for the agreement were directed to the Navajo Nation's Division of Natural Resources. The National Environmental Policy Act (NEPA) compliance documentation (which includes the Environmental Assessments as well as the Remedial Action Plans) was developed with the participation of the Navajo Nation. The characterization of surface and subsurface radiological contamination involved a phased approach that included an aerial assessment, a mobile scan, and on-site sampling to delineate the extent of contamination according to specific removal criteria set by the EPA. Over $140 million was spent on the surface remediation of the four uranium mill processing sites and construction of three disposal cells. That agreement expired in 1998 along with the surface cleanup authority, and a new cooperative agreement was initiated for groundwater cleanup that remains in place. The cooperative agreement for groundwater was recently extended for another five years through March 2017. However, DOE will continue groundwater monitoring and/or remediation activities until the sites are in compliance with the EPA standards.

DOE funds maintenance of groundwater remediation activities and LTS&M at the four Navajo Nation sites for a cost over $4 million dollars annually. The only UMTRCA sites with active remediation systems are located on Navajo Nation: Shiprock and Tuba City.

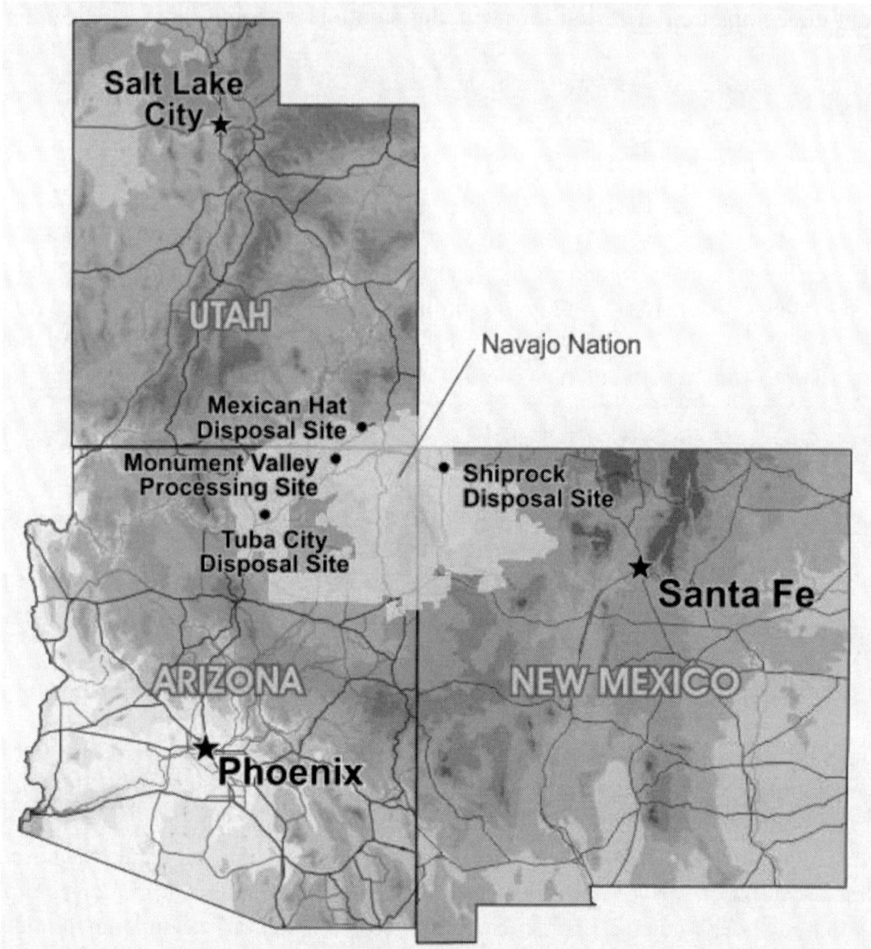

Figure 1. Location of UMTRCA Sites on Navajo Nation.

The groundwater compliance strategies are reviewed annually with the Navajo Nation and the NRC to track progress towards cleanup standards in the groundwater plumes. DOE will continue to work with the Navajo Nation to complete the groundwater remediation responsibilities.

Also, DOE is continuing studies of the sources of constituents in Many Devils Wash near the Shiprock New Mexico Disposal Cell and Processing Site. This work is part of a broader effort by DOE to understand sources of naturally occurring contaminants in the region. For example, at Many Devils Wash, DOE has published several reports that together provide strong evidence that some constituents (e.g. uranium, selenium, nitrate, and sulfate) in the wash are naturally occurring and not a result of ore processing. DOE's studies have shown that these constituents are found naturally in the Mancos Shale, a Cretaceous marine deposit widely occurring in the southwestern United States. The studies included sites with no known uranium mining or milling, and therefore, the elevated levels of uranium, selenium, nitrate and sulfate were not due to uranium extraction processes. These sites are termed "analog sites".

Links to DOE's reports are provided:

- *Application of Environmental Isotopes to the Evaluation of the Origin of Contamination in a Desert Arroyo: Many Devils Wash, Shiprock, New Mexico,* can be found at http://www.lm.doe.gov/WorkArea/linkit.aspx?LinkIdentifier=id&ItemID =9505&libID=9611.
- *Naturally Occurring Contaminants from the Mancos Shale*, can be found at http://www.lm.doe.gov/WorkArea/linkit.aspx?Link Identifier=id&ItemID=7924&libID=8030 , and
- *Geology and Groundwater Investigation Many Devils Wash, Shiprock Site, New Mexico*, can be found at http://www.lm.doe.gov/WorkArea/linkit.aspx?LinkIdentifier =id&ItemID=7743&libID=7850.

Five-Year Plan Goal

"Remediate Groundwater at the Inactive Uranium Milling Sites: DOE's Office of Legacy Management will continue maintenance of existing groundwater treatment activities at the three inactive uranium ore processing sites and long-term surveillance and maintenance at four inactive milling sites, all of which have been remediated by DOE Environmental Management throughout the 5-year period of this plan. Site surveillance and maintenance costs is estimated at $4,940,000 in 2008 and $3,178,000 in 2009".

Summary of Actions

DOE's Office of Legacy Management, in consultation with the Navajo Nation, and with NRC's regulatory oversight, has been addressing groundwater contamination at the four sites. Note that at the Mexican Hat, Arizona site, groundwater contamination is confined and the hydraulic conditions prevent the future use of any groundwater within the area; therefore, DOE and the Navajo Nation have agreed that no further remedial action is warranted at the site. As required by UMTRCA, DOE has worked closely with the Navajo Nation Division of Natural Resources, and with other federal and state agencies in the cleanup of uranium milling sites.

Milling Site Action Plan

- DOE will continue to monitor the three disposal cells to ensure they remain effective in protecting human health and the environment.
- DOE funds maintenance of groundwater remediation activities and LTS&M at the four Navajo Nation sites at a cost of about $4 million dollars annually[2]. The groundwater compliance strategies are reviewed annually with the Navajo Nation and the NRC to track progress towards cleanup standards in the groundwater plumes.

DOE will continue to work with the Navajo Nation using the currently approved plans to complete our groundwater remediation responsibilities.

- DOE will continue to fund the Navajo Nation, under the Cooperative Agreement, to assure they have the resources to review and participate in DOE's activities as we continue these long term actions.
- DOE will continue its evaluation of naturally occurring groundwater constituents, such as those at Many Devils Wash, which may have been erroneously attributed to uranium milling. If such evidence continues to be supported by research results from DOE and other agencies, DOE may reevaluate the need for some water treatment efforts at mill tailing sites on the Navajo Nation.

DOE continues to collaborate with the other federal agencies and the Navajo Nation on issues on the four Navajo sites. DOE activities have included:

- Attending annual meetings and working with other stakeholders (DOE, EPA, IHS, BIA, DOI, NRC, ATSDR, Navajo and Hopi Tribes, and the public) on uranium contamination issues and updates on each of the federal agency's progress (November 2011 and March 2012); Holding quarterly meetings with the Navajo Nation to discuss groundwater data and strategies;
- Meeting with stakeholders in Shiprock in March 2012 to discuss DOE Legacy Management program's LTS&M activities at the Shiprock site and to address concerns raised by members of the community;
- Meeting with Navajo Nation and the United States Geological Survey (USGS) in September 2012 to discuss LM's geochemistry studies performed on the groundwater at the Shiprock site and to discuss USGS's future plans for groundwater studies at the site;
- Working closely with NNEPA staff on supporting the remediation of the Highway 160 site, providing technical assistance throughout the remediation, and operating the Grand Junction Disposal Site to receive the contaminated material;
- Providing groundwater monitoring data, annual treatment system performance reports, and annual site inspection reports on LM website so it is available to the public; and
- Supporting summer internships for Navajo Nation students at the DOE Grand Junction Office to help build additional tribal capacity in environmental fields.

The NRC's regulatory program is based on UMTRCA and governs the management and remediation of legacy mill sites by the DOE and the licensing of new uranium recovery facilities that are owned by private entities. The four Navajo Nation legacy sites addressed under Title I of UMTRCA are under the general license issued by the NRC pursuant to 10 CFR 40.27 with DOE as the general licensee. The purpose of the general license is to ensure that the tailings disposal sites are cared for in a manner that protects public health and safety and the environment once the remedial actions are completed. The NRC oversees DOE activities (as the licensee) at these sites and concurs in remedial action plans, reviews and comments on DOE environmental and performance reports, and conducts DOE inspection activity oversight.

During the 1980s after the issuance of UMTRCA, the NRC implemented many regulatory improvements to avoid the problems from uranium milling operations conducted in the 1950s and 1960s. Some of these are:

- Implementation of below grade disposal of mill tailings;
- Use of liners for tailings impoundments to prevent groundwater contamination;
- Installation of groundwater monitoring wells to monitor potential groundwater contamination;
- Financial surety requirements to reduce the risk of problems caused by a licensee's bankruptcy; and,
- Reclamation and decommissioning requirements.

The NRC's uranium recovery program regulatory framework is designed to adequately protect public health and safety and the environment. Currently there are no new sites planned on the Navajo Nation. One site, the Hydro Resources, Inc. (HRI) facility, is in Crownpoint, NM, near the Navajo Reservation and in the Church Rock area adjacent to Indian Trust Lands. The Church Rock mill site is adjacent to the Navajo Reservation.

At the request of the Navajo Nation, NRC staff has begun Government-to-Government meetings with the Navajo Nation regarding the HRI site. The NRC is committed to keeping the Navajo Nation informed of all meetings with applicants for uranium recovery sites at or near Navajo land so that they may participate in the meetings.

Objective 8. Case Control Studies of Health Risks Faced by Individuals Residing Near Mill Sites or Abandoned Mine Sites

Through Congressional authorization, several efforts were undertaken in the past to compensate uranium mine and mill workers and Navajo Nation residents for radiation exposure.

Radiation Exposure Compensation Act (RECA)

The U.S. conducted nearly 200 atmospheric nuclear weapons development tests from 1945 to 1962. Essential to the development of nuclear weapons was uranium mining and processing, which was carried out by tens of thousands of workers. Congress passed the RECA (42 U.S.C. note §2210) in 1990 to provide partial restitution to individuals who developed serious illnesses after exposure to radiation released during atmospheric nuclear testing or following employment in the uranium industry. Residents from the Four Corners region of Utah, Colorado, Arizona, and New Mexico, have filed the majority of claims. Approximately 10% of all claimants are members of one of several Native American tribes. The Navajo Nation maintains one office which assists potential claimants through the Office of Navajo Uranium Workers program. The RECA program is administered by the Department of Justice. RECA provides for payments of $100,000 for uranium miners, millers, and ore transporters, \$75,000 for individuals who participated onsite during atmospheric nuclear weapons testing, and \$50,000 for individuals who lived or worked downwind of the atmospheric tests. The federal government has approved over 26,000 claims and awarded over \$1.7 billion under the RECA (as of 1/02/2013).

Energy Employees Occupational Illness Compensation Program Act (EEOICPA)

EEOICPA (codified at 42 U.S.C. §§7384d-7384g) was enacted by Congress in 2000 to provide lump-sum compensation and health benefits to eligible (DOE) nuclear weapons workers (including employees, former employees, contractors and subcontractors) and lump-sum compensation to certain survivors if the worker is deceased. The EEOICPA program is now administered by the Department of Labor (DOL), with cooperation from DOE and HHS.

The DOL is supported in its role by the DOE Office of Health, Safety and Security, and the Department of Justice. Part E of the Act also covered exposure to toxic substances, which included not just radiation, but chemicals, solvents, acids and metals. To date the federal government has paid workers and certain survivors over \$8.8 billion.

For covered employees, a maximum of \$400,000 of compensation has been made available to employees with covered illnesses related to the production of weapons related materials, recently including uranium mill cleanup contractor employees such as those involved with the four former uranium processing sites in the Navajo Nation.

Indian Health Service Treatment, Public Health, and Research Activities

Five-Year Plan Goal

"The Indian Health Service will continue to diagnose and treat known health conditions in eligible Indians, and also support a university-led Navajo Uranium Assessment and Kidney Federal Actions to Address Impacts of Uranium Contamination in the Navajo Nation Five-Year Plan Summary Report Health Project in 2008. IHS will also review existing databases to develop plans for improved cancer case surveillance, review water contamination data for potential future health studies, and develop plans to assess the prevalence of cancer and other health conditions for populations near AUM and inactive mill sites. Cost estimates for this objective are unavailable, but are projected to be achievable within current and projected agency appropriations."

Summary of Actions

IHS, within the Department of Health and Human Services (HHS), is an Agency partner contributing over the years to the Navajo Uranium Five-Year Plan. NAIHS, located in Window Rock, Arizona, is one of 12 service areas of IHS. The major funding source for NAIHS programs and services is from Congressional appropriations. Additional sources of funding for environmental health services include resources from the EPA and the Navajo Nation. IHS identified personal and environmental health-related objectives and progress against three (3) major objectives as reported in this section of the report.

IHS Objective #1 – Diagnose and Treat Health Conditions for Eligible IHS Beneficiaries

As noted earlier, the IHS is primarily a healthcare service delivery Agency serving eligible American Indian beneficiaries. During the course of the Five-Year Plan, sub-objectives were developed and targeted to segments of the service population and healthcare providers as follows.

A. Training to healthcare providers
   In 2009 a CDC-ATSDR consultant provided training at five NAIHS hospitals and clinics to over 100 health professional staff across the Navajo reservation. The training covered the physics, exposure routes and the medical impact of exposure on humans for uranium as well as other heavy metals. In 2010 this type of training was provided once again in Window Rock, Arizona, utilizing CDC- ATSDR, Northern Arizona University, University of New Mexico and NAIHS staff as presenters targeted to over 130 Navajo Nation paraprofessionals as well as health care professionals.
B. Radiation Exposure Screening and Education Program (RESEP)
   The NAIHS has deployed, since 1990, a service delivery model for a screening and education program, most recently with grant resources from HHS through its Health Resources and Services Administration (HRSA) ($200,000 in 2012) targeted towards potential Radiation Exposure and Compensation Act-eligible, occupationally exposed individuals, including former miners, millers and others. Program staff provided medical screening exams, education, and case management services at multiple Navajo Area hospital and health care locations to individuals exposed to uranium as a result of many years of work in the uranium mining industry on and off the Navajo reservation. In addition to the HRSA grant funds, NAIHS has spent an estimated $257,000 of appropriated funds annually for laboratory tests, X-Rays, and office visit costs to supplement the efforts of the RESEP grant funded staff salaries and benefits.
C. Community Uranium Exposures Journey to Healing Program
   Based upon the RESEP experience and in conversations with community leaders and residents the NAIHS created in January 2010 the Community Uranium Exposure Journey to Healing (CUE-JTH) program, which provides the following:
   1. Community forum conversations providing health education to individuals across the reservation and gathering information on people's individual concerns and needs regarding their health problems that they believe may be attributable to non-occupational exposure to uranium.
   2. Health screening and medical monitoring/case management services in clinics as well as in local community settings, utilizing at times a NAIHS mobile clinic vehicle.
   3. Continuing education to hospital- and clinic-based health care professional and paraprofessional staff across the reservation.

Congressionally appropriated IHS Hospitals and Clinics funds ($175,000 in 2009, $175,000 in 2010, $300,000 in 2011, and $300,000 in 2012) were identified by the NAIHS

Director to support the CUE-JTH program to cover the salaries of the current staff (3.5 FTEs) and operational costs.

Since the first employee began CUE-JTH work in 2010, community conversations have been held in 14 locations with over 200 participants. Contaminated water has easily been identified as the leading cause of concern among residents attending each community conversation.

Summary of Results

Twenty-two community-based medical screening exam events were conducted by CUE-JTH staff in 2011-2012. During these events 699 individuals presented themselves for examinations, and 578 of these individuals self-identified current or past non-occupational exposure to uranium. Results of the events are summarized below in Tables 7 and 8.

Processes to share CUE-JTH information with each person's primary care provider (system in place) and in the future with the NNEPA (planned) and the NNDOH (referral system design under discussion) are part of CUE-JTH goals. Health information results were shared with the University of New Mexico kidney health research staff for those individuals enrolled in their NIH funded study (described below under IHS Objective 3).

Table 7. Exposure Routes Identified by Community Members

Route	Current	Past
Water	44	277
Abandoned Mines	142	123
Buildings	39	135
Meat	88	217
Plants	72	148

Table 8. Frequent Health Issues Identified by Community Members

Health Issue	Individuals	Health Issue	Individuals
Diabetes	201	Anxiety	117
Headaches	224	Skin	102
High Blood Pressure	253	Miscarriages	68
Eye Problems	289	Cancer	57
Arthritis	183	Developmental Issues	32
Depression	157	Learning Issues	72
Respiratory	131	Kidney	77
Heart	86	Speech Delay	63

In addition to the above activities NAIHS CUE-JTH program staff will record medical screening and monitoring information in a database within the NAIHS patient record system for use by case management staff and potentially for future research activities. (It should be noted that there has not been any determination that any of these health issues is linked to uranium exposure.)

IHS Objective #2—Environmental and Public Health Work in Cooperation with Other Agencies

As part of IHS staff work, the IHS Director has emphasized partnering with other Agencies and organizations to support the IHS mission. Two major sub-objectives consistent with the Five- Year Plan were undertaken: fencing of the Tuba City Landfill and the Provision of Safe Water/Sewers to Homes, discussed in more detail in objectives 6 and 2, respectively, above.

IHS Objective #3—Conduct Activities Supporting Health Research on Non-Occupational Exposure to Uranium

The NAIHS in the provision of direct healthcare services actively supports Navajo Nation Institutional Review Board approved research activities. During the Five-Year Plan timeframe, work was conducted by IHS staff on research activities related to non-occupational exposure to uranium follows.

A. Navajo Uranium Assessment and Kidney Health Project:
   The NIH awarded the above project to the University of New Mexico. From 2011-2012, NAIHS CUE-JTH staff assisted UNM staff by completing medical evaluations on enrolled study participants living in the New Mexico portion of the reservation. NAIHS staff from the Crownpoint, New Mexico, Service Unit provided other in-kind services to the researchers as part of an agreement entered into between NAIHS and UNM.
   To date all articles resulting from this study are in the pre-publication phase of work by UNM staff. In community presentations and in communications with NAIHS staff, an increase in kidney disease among study participants seemed to be related more to a history of active mining rather than non-occupational uranium exposure.
B. Assistance to ATSDR Prospective Birth Cohort Study Involving Environmental Uranium Exposure on the Navajo Nation:
   This CDC-ATSDR study, including the assistance by NAIHS to the study, is discussed below.

Future Plans

The NAIHS intends to continue work within the framework of the three (3) objectives noted earlier in this report. Training for health professionals and paraprofessionals on the occupational and non-occupational health impact of uranium on beneficiaries will continue. The NAIHS medical monitoring program for beneficiaries will continue to place an increased emphasis on screening individuals referred by Navajo Nation and other Federal agency partners along with the expansion of the database of information of screening results for potential use by future researchers. Completion of ongoing safe water projects will occur in the period covered by the next Five-Year Plan with new projects designed and constructed once information has been developed on their feasibility. Finally, NAIHS will collaborate with health researchers and the NNDOH to complete a prospective birth cohort study on the potential impact of uranium exposure on pregnant women and infant health.

ATSDR Prospective Birth Cohort Study Involving Environmental Uranium Exposure in Navajo Nation

Study Objectives

To evaluate the potential association between environmental contaminants (i.e., uranium and other heavy metal exposures) and reproductive birth outcomes, by recruiting Navajo mothers, assessing their uranium exposure at key developmental milestones, and then following their children post-birth to evaluate any associations with birth defects or developmental delays. Exposures will be assessed through bio-monitoring, home assessments, and surveys. The information generated by this study may be of value in developing programs to mitigate environmental uranium exposure. Additionally, the study will provide additional baseline data on exposures, behaviors, and chronic health that may be of value in developing health education and outreach to increase prenatal care utilization.

Summary of Actions

In 2009, ATSDR representatives met with Navajo Nation Division of Health (NNDOH), NNEPA, NAIHS, EPA, and local university researchers to gain further understanding of previous research activities conducted at the Navajo Nation, to summarize current activities, and to discuss knowledge gaps in environmental uranium exposure and potential health effects. In addition, NNEPA staff led ATSDR representatives on a tour through parts of the reservation to observe some of the abandoned mine areas and three of the four milling sites. While on the tour, ATSDR staff had the opportunity to discuss health concerns with community members.

In 2010, ATSDR published a Funding Opportunity Announcement, "A Prospective Birth Cohort Study Involving Environmental Uranium Exposure in the Navajo Nation," soliciting investigator initiated proposals. In August 2010, an Interagency Agreement (IAA) with NAIHS was finalized to support the Cooperative Research Agreement. The IAA outlined provisions for NAIHS's role in facilitating the clinical and recruitment/outreach (through Navajo Nation Division of Health) aspects of the overall study.

The Cooperative Research Agreement for the Prospective Birth Cohort Study was awarded to Dr. Johnnye Lewis, Director of the Community Environmental Health Program at the University of New Mexico (9/21/10 -8/31/13). ATSDR representatives presented the study design and objectives of the Prospective Birth Cohort Study at the 2010 Uranium Contamination Stakeholder Workshop in Tuba City, Arizona and addressed questions from the Navajo community.

In October 2010, ATSDR and UNM hosted the Navajo Prospective Birth Cohort Study Kick-off meeting at the UNM Center for Development and Disability in Albuquerque. Attendees included UNM staff, NAIHS, NNDOH, NNEPA, Community Liaison Group, and Growing in Beauty. The purpose of the meeting was to discuss stakeholder roles and responsibilities and other issues that needed to be addressed before commencing the study.

In April 2011, ATSDR presented the study objectives with UNM to Navajo Nation Human Health and Review Board (NNHHRB) during the HHS Tribal Consultation Meeting in Window Rock, AZ. That August, ATSDR awarded a sole source contract to the NNDOH to provide health education, training, and outreach regarding the study. NNHRRB and University of New Mexico Institutional Review Board (UNMIRB) approved the study protocol. CDC/ATSDR formalized an agreement to rely on UNMIRB for approval. ATSDR presented at the Uranium Contamination Stakeholder Workshop in which was held in Farmington, NM, November 8-10. The 60-Day Federal Register Notice for the study was published on November 22, 2011. No comments were received.

In January 2012, ATSDR provided study updates at the Tribal Advisory Committee Meeting in Atlanta, GA. During February and March, ATSDR assisted with survey administration and bio- monitoring training of research study team. The 30-Day Federal Register Notice of the study was published on March 27, 2012. No comments were received. In May 2012, ATSDR, UNM, and NNDOH presented the Navajo Birth Cohort study at the National Indian Health Board 2012 Tribal Public Health Summit in Tulsa, OK.

UNM, NDOH, and IHS are conducting trainings among the study team in order to successfully recruit approximately 1,500 pregnant participants once field work begins. UNM has also conducted several laboratory trainings with the IHS hospitals to ensure proper collection, processing, and shipment of study samples. When the study is approved, NAIHS and Pub. L. No. 93-638-contracted and -compacted staff will collect specimens from pregnant women who enroll in the study and from their children who are enrolled after their births.

AGENCY COLLABORATION AND COMMUNITY ENGAGEMENT

Significant and active agency collaboration between federal, tribal, and academic partners was a key part of achieving success during the five-year period. As discussed above, multiple agencies worked together on nearly every aspect of the Plan.

The following projects and events highlight some of the ways various groups worked together to achieve the goal of providing opportunities for meaningful community engagement.

- Annual Uranium Contamination Stakeholder Workshops (UCSW) occurred in 2008, 2009, 2010, and 2011. The UCSW provides a forum for federal agencies to discuss their goals, programs and activities for the Five-Year Plan and get feedback from stakeholders. In addition, the Uranium Contamination Stakeholders Workshop provides education and facilitated discussions. Due to the release of the Five-Year Plan Report (2008-2012) and the need for community input on the completed Five-Year Plan Report (2008-2012) and the next Five-Year Plan (2013-2017), the next Uranium Contamination Stakeholder Workshop will take place in April 2013.
- The EPA, HUD and IHS partnered to provide $26.7 million for 14 projects that will provide piped water to 808 homes.
- DOE and NNEPA staff worked together to complete remediation of the Highway 160 site. DOE provided funding and technical assistance to NNEPA to conduct the remediation and for ultimate disposal of waste at the DOE Grand Junction Disposal

Site. The NRC reviewed the project completion report documenting the site cleanup and agreed that the site met the 40 CFR 192 cleanup standards.

- The IHS Community Uranium Exposure-Journey to Healing program organizes health fairs and health screenings at many Navajo Nation Chapter Houses around the Navajo Nation. EPA, NNEPA and other Navajo departments have participated in health screening events in order to provide information about soil and water contamination due to uranium.
- Representatives from the EPA, New Mexico Environment Department, the Navajo Nation, DOE and the NRC have worked together since 2011 to dis cuss and resolve technical, regulatory and policy issues related to the cell pre-design for the NECR mine waste repository.
- BIA, the EPA, the Hopi Tribe and the Navajo Nation have collaborated during investigation at the Tuba City Dump. As part of this collaboration, BIA provided funding to the Hopi Tribe to conduct a Wellhead Protection Study.
- The EPA partnered with Navajo Nation to hold the first tribal Superfund Job Training Initiative (Super JTI) program. The Superfund Job Training Initiative is a national job training program that provides free technical training, job skills, and networking opportunities to program participants. The EPA worked with SuperJTI program participants to empower and prepare them for job opportunities. The 20 program graduates received certificates in CPR/First Aid, Radon Measurement and Mitigation, and Hazardous Waste Operations and Emergency Response (40 Hour HAZWOPER) at a graduation ceremony on December 4, 2012.

SUMMARY OF AGENCY EXPENDITURES

Since 2008, Congressional appropriations have supported the activities related to the Five-Year Plan to address uranium contamination on the Navajo Nation. Agencies have spent close to $110 million conducting these activities. The following tables provide actual extramural expenditures data by agency.

In addition to the amounts spent from the EPA's budget, Responsible Parties have contributed over $17 million to the investigation and cleanup of abandoned mines either directly or through EPA use of funds recovered in bankruptcy from Tronox, Inc., formerly Kerr-McGee.

Summary of EPA Expenditures (in millions)

Expenditure	2008	2009	2010	2011	2012	Total
Structures: Assessment and Remediation	$3.4	$5.3	$1.3	$5.3	$5.4	$20.7
Water: Source Contamination Assessment	$0.3	$6.3	$2.4	$0.3	$0.7	$9.9
Abandoned Mines: Assessment and Cleanup	$0.3	$1.8	$7.3	$3.8	$0.8	$14.0
Grants to Navajo Nation	$1.3	$1.5	$1.5	$1.3	$0.7	$6.3
Total Expenditures (EPA Budget)	**$5.3**	**$14.9**	**$12.5**	**$10. 7**	**$7.6**	**$50.9**

Summary of BIA Expenditures (in millions)

Expenditure	2008	2009	2010	2011	2012	Total
Tribal support	$0.08	$0.39	--	--	$0.15	$0.62
Studies and IAR	--	--	$0.01	--	--	$0.01
EPA/Hot spot study	$1.69	--	--	$0.30	$0.15	$2.14
Interim Actions	$0.09	--	--	--	--	$0.09
Consultation	$0.11	--	--	--	--	$0.11
RI/FS	$0.16	$0.24	$1.80	$0.14	$1.44	$3.78
Groundwater	$0.09	$0.19	$0.44	$0.12	$0.04	$0.88
Total Expenditures (BIA Budget)	**$2.2**	**$0.8**	**$2.3**	**$0.6**	**$1.8**	**$7.6**

Summary of DOE Expenditures (in millions)

Expenditure	2008	2009	2010	2011	2012	Total
Mexican Hat	$0.02	$0.02	$0.02	$0.03	$0.02	$0.11
Monument Valley	$0.43	$0.65	$0.25	$0.47	$0.43	$2.23
Shiprock	$0.56	$0.91	$0.91	$1.13	$1.27	$4.78
Tuba City	$2.9	$2.6	$2.42	$2.82	$3.22	$13.96
Highway 160 Site Clean up	--	--	--	$5.00[3]	$1.00	$6.0
Navajo Nation Cooperative Agreement	$0.47	$0.24	$0.33	$0.55	$0.45	$2.04
Hopi Indian Tribe Cooperative Agreement[4]	$0.10	$0.04	$0.03	$0.06	$0.03	$0.26
Total Expenditures (DOE Budget)	**$4.5**	**$4.5**	**$3.9**	**$10.1**	**$6.4**	**$29.4**

Summary of IHS Expenditures (in millions)

Expenditure	2008	2009	2010	2011	2012	Total
RESEP Costs[5]	$0.3	$0.3	$0.3	$0.3	$0.3	$1.5
Medical Monitoring[6]	--	$0.2	$0.3	$0.3	$0.3	$1.1
Tuba City[7]	--	$0.1	--	--	--	$0.1
Water Projects[8]	--	$6.6	$3.7	$2.5	$0.3	$13.1
Kidney Health Project[9]	--	--	--	$0.1	$0.1	$0.2
Total Expenditures (IHS Budget)	**$0.3**	**$7.2**	**$4.3**	**$3.2**	**$1.0**	**$16.0**

Summary of ATSDR Expenditures (in millions)

Expenditure	2008	2009	2010	2011	2012	Total
Cooperative Agreement with University of New Mexico for Prospective Study (Birth Cohort Study)	--	--	$1.0	$1.0	$1.0	$3.0
NNDOH for Health Education, Training & Outreach	--	--	$.4	$.6	--	$1.0
Inter-Agency Agreement withIndian Health Service for Medical Screenings, Prenatal Care, & Administrative Activities	--	--	$.4	$.4	$.4	$1.1
Biomonitoring and LaboratoryExpenses	--	--	--	--	$.4	$.4
Total Expenditures (ATSDR Budget)	**--**	**--**	**$1.8**	**$2.0**	**$1.8**	**$5.5**

NRC Expenditures

Resources associated with the Five-Year Plan are included in NRC's Decommissioning and Low-Level Waste Program. NRC does not currently have any contracts for cleanup or oversight of the sites on the Navajo Nation so resources associated with the Program are primarily staff activities. To date, the NRC has expended approximately 830 staff hours in activities directly related to the Five-Year Plan. The Commission has provided the staff with adequate resources to conduct oversight of UMTRCA sites, to perform safety reviews for new uranium recovery applications, and to maintain outreach with the Navajo Nation, and it is anticipated that the Commission will provide the staff with the same level of resources in the foreseeable future.

Future Expenditures

The funding for the second Five-Year Plan cannot be reasonably estimated at this time as much of the follow-on work is dependent upon on-going studies and assessments. The agencies will continue to submit resource requirements on an annual basis through the budget formulation process for Congressional consideration. Congressionally appropriated resources will then be allocated by agencies to projects and activities based upon future project plans and program funding prioritization needs and guidelines including, for example, EPA's response program's "worst sites first" principle.

As discussed in Objective 4 above, funding provided by PRPs will supplement Agency resources to address uranium contamination. GE/UNC has agreed to pay past costs and costs of a 2012 interim removal at the NECR Site. Region 9 and Region 6 anticipate proposing additional agreements with GE/UNC to fund the proposed long-term remedial action for the NECR Site, which is anticipated to cost approximately $44 million, following Region 6's issuance of a Record of Decision, scheduled for early 2013. As noted above, EPA is negotiating agreements with El Paso (subsidiary of Kinder Morgan) to perform response actions in the Cameron area and with Western Nuclear, Inc. (subsidiary of Freeport-McMoRan) to perform actions at the Ruby Mines.

Conclusion

The agencies met the overarching goal of protecting Navajo residents from urgent risks first by addressing structures, water supplies, mills, dumps, and mines with the highest levels of radiation. Much more was learned about the scope of the problem and it is clear that additional work will be needed. All agencies are committed to continue working with the Navajo Nation to reduce the health and environmental risks and to finding long-term solutions to the remaining uranium issues on Navajo lands.

REFERENCES

Markey, E. 2012. Letter from E. Markey, H. Waxman, B. Lujan, F. Pallone, R. Grijalva, and M. Heinrich, Congressmen, to Secretary of Energy S. Chu regarding Navajo Nation Five-Year Plan, October 18.

DOL. (U.S. Department of Labor), DEEOIC (Division of Energy Employees Occupational Illness Compensation). "Program Benefits," Office of Workers' Compensation Programs, http://www.dol.gov/owcp/energy/regs/compliance/progbenefits.htm, accessed 20 December 2012.

DOL. (U.S. Department of Labor), EEOICP (Energy Employees Occupational Illness Compensation Program), http://www.dol.gov/owcp/energy/index.htm, accessed 20 December 2012.

EPA. (U.S. Environmental Protection Agency), 2011 Presentations webpage. "Addressing Uranium Contamination in the Navajo Nation," EPA Region 9, http://www.epa.gov/region09/superfund/navajo-nation/stakeholder.html, accessed 20 December 2012.

EPA. (U.S. Environmental Protection Agency), *Comprehensive Database and Atlas*. EPA Region 9 http://yosemite.epa.gov/r9/sfund/r9sfdocw.nsf/3dc283e6c5d6056f8825742 6007417a2/8ed3f74d2 f55b845882573f400787a8e!opendocument, accessed 20 December 2012.

EPA. (U.S. Environmental Protection Agency), *Uranium Location Database Compilation*, Office of Radiation and Indoor Air, Radiation Protection Division, EPA 402-R-05-009, August 2006.

EPA. (U.S. Environmental Protection Agency), Tuba City Open Dump http://yosemite.epa.gov/r9/sfund/r9sfdocw.nsf/3dec8ba3252368428825742600743733/2a32894331dfa3b888 25771200761e61!OpenDocument#documents, accessed 20 December 2012.

EPA 1971 (U.S. Environmental Protection Agency), Progress Report on Uranium Mill Tailings Activities, 1971. R.J. Augustine. December 31.

DOE 1999 (U.S. Department of Energy), *Uranium Mill Tailings Remedial Action Surface Project 1979-1999, End-of-Project Report*, DOE/AL/62350-50020, May.

DOE 2012 (U.S. Department of Energy), *Multivariate Statistical Analysis of Water Chemistry in Evaluating the Origin of Contamination in Many Devils Wash, Shiprock, New Mexico,*LMS/SHP/S09257 ESL-RPT-2012-03, Office of legacy Management, Grand Junction, CO, December.

House Report 2007(U.S. House of Representatives), "Hearing On The Health And Environmental Impacts Of Uranium Contamination In The Navajo Nation", House of Representatives, Committee on Oversight and Government Reform, Washington, D.C. October 23.

NNAML (Navajo Nation, Division of Natural Resources Abandoned Mined Lands), Navajo Nation AML Reclamation/UMTRA Department webpage http://www.aml.navajo-nsn.gov/AML_Files/AMLReclamationPage.html, and Federal Actions to Address Impacts of Uranium Contamination in the Navajo Nation Five-Year Plan Summary Report http://www.aml.navajo-nsn.gov/AML_Files/AMLAccomplishmentsPage.html, accessed 20 December 2012.

Navajo Nation, *Groundwater Investigation Report UMTRA DOE Office of Legacy Management Site, Tuba City, Coconino County, Arizona*, 2011.

End Notes

[1] As defined in "Health and Environmental Impact of Uranium Mining in the Navajo Nation Five-Year Plan", 2008

[2] For comparison, DOE spends an average of $200,000 per year per site on its other UMTRCA sites.

[3] $5 million was appropriated by Congress under the Omnibus Appropriations Act, 2009 (Pub. L. No. 111-8) for the Highway 160 site; Navajo Nation expended the majority of funds, leading the cleanup in 2011 and early 2012 through a cooperative agreement with DOE. An additional $1 million was spent by DOE to remove the waste material, transport it, and place it in the Grand Junction, CO disposal cell, the only UMTRCA site that remains open. The cost distribution between 2011 and 2012 is an estimate.

[4] The cooperative agreement provides support for the Hopi Tribe for UMTRCA disposal sites located on the Navajo Nation which are near Hopi lands (e.g., Tuba City, AZ).

[5] RESEP (Radiation Exposure Screening and Education Program) costs incurred by IHS supplementing HHS HRSA grant funds.

[6] Medical Monitoring Community Outreach and Medical Screening Program costs.

[7] Tuba City Landfill fencing project.

[8] IHS-only costs for water projects for homes within a 10-mile radius of abandoned mines or contaminated stock wells. Excludes EPA and Navajo Nation contributed funds.

[9] Estimated IHS in-kind contributions to University of New Mexico Kidney Health Project.

INDEX

A

abandoned mines, vii, 1, 3, 4, 11, 14, 17, 22, 32, 33, 34, 35, 36, 37, 38, 39, 46, 47, 51, 53, 69, 71, 74, 82, 108, 113, 117, 126, 144, 148
abandoned uranium mines, vii, 2, 4, 6, 11, 13, 14, 15, 18, 19, 23, 24, 31, 34, 35, 36, 38, 40, 43, 52, 64, 66, 69, 70, 83, 102, 104, 109, 112, 117
abatement, 78
access, 21, 35, 42, 43, 45, 67, 71, 77, 81, 93, 112, 117, 118, 122
accommodations, 76
accountability, 68
accounting, 31, 48
acquisitions, 29
agency collaboration, 106, 143
American Recovery and Reinvestment Act of 2009, 23
AOC, 105, 132
appropriations, 22, 23, 25, 64, 73, 138, 144
aquifers, 63
arsenic, viii, 78, 79, 101, 112, 130
assessment, 2, 5, 15, 17, 18, 20, 23, 26, 28, 29, 30, 33, 38, 39, 40, 46, 48, 50, 51, 53, 58, 59, 60, 61, 62, 65, 68, 69, 70, 71, 72, 74, 75, 76, 77, 82, 83, 85, 91, 94, 103, 105, 106, 107, 117, 118, 130, 131, 133
attribution, 114
audit, 6, 54
AUMs, vii, 69, 75, 81, 82, 83, 117
authorities, 32, 73, 91, 93, 95, 130
authority, 12, 13, 18, 40, 64, 68, 71, 72, 75, 76, 79, 83, 84, 85, 89, 90, 109, 112, 128, 129, 133
awareness, 44

B

background radiation, 11, 17
bacteria, 37, 67, 78
bankruptcy, 24, 33, 65, 98, 127, 137, 144
base, 33, 52
base year, 52
behaviors, 142
benchmarks, 32
beneficiaries, 139, 141
benefits, 18, 22, 65, 104, 113, 138, 139
BIA, 2, 3, 5, 10, 15, 19, 20, 25, 26, 27, 28, 29, 30, 34, 39, 48, 49, 50, 52, 53, 54, 56, 59, 63, 65, 66, 70, 72, 82, 86, 87, 89, 90, 91, 92, 93, 94, 96, 97, 102, 105, 128, 130, 131, 132, 136, 144, 145
bicarbonate, 63
births, 20, 143
bonding, 10
bone, viii, 3, 101, 119
bone cancer, viii, 3, 101, 119
bonuses, 10
boreholes, 92
breakdown, 54, 57, 59, 95
Bureau of Indian Affairs (BIA), v, 2, 3, 4, 15, 19, 25, 51, 54, 56, 69, 72, 90, 101, 102, 130
by-products, 10

C

cancer, 11, 73, 96, 97, 105, 138
candidates, 72, 82, 126
capacity building, 22, 46, 109
carbon, 63
carbon dioxide, 63
carcinogen, 75, 108
career development, 22
CDC, 3, 11, 16, 19, 20, 46, 47, 52, 63, 67, 79, 97, 103, 105, 112, 113, 139, 141, 143
Census, 6, 7
CERCLA, 3, 11, 12, 15, 16, 32, 39, 50, 63, 64, 81, 82, 83, 94, 106, 119, 122, 127, 131

CFR, 94, 98, 99, 105, 130, 133, 136, 144
challenges, vii, 1, 2, 5, 28, 38, 40, 43, 48, 52, 53, 83
chemical(s), 64, 72, 79, 90, 94, 129, 138
children, 6, 11, 36, 142, 143
citizens, 68
cleaning, 26, 27, 32, 41, 50, 62
closure, 15, 74, 83, 91, 93, 94, 131, 132
clothing, 75, 108
clusters, 6, 92
coal, 64
collaboration, 2, 38, 45, 46, 48, 50, 68, 82, 83, 103, 107, 144
college students, 22
commercial, 10
communication, 26, 28, 50
communities, 2, 3, 5, 9, 11, 15, 16, 19, 20, 21, 22, 23, 30, 32, 36, 38, 39, 40, 41, 42, 43, 44, 45, 47, 52, 53, 62, 64, 67, 68, 71, 75, 81, 90, 96, 107, 112, 117, 130
community, 4, 6, 11, 16, 18, 19, 21, 22, 26, 38, 39, 40, 41, 42, 43, 44, 45, 46, 48, 50, 53, 76, 77, 94, 95, 104, 105, 106, 107, 118, 121, 122, 136, 139, 140, 141, 142, 143
compensation, 20, 62, 75, 77, 103, 138
competing interests, 48
compilation, 5
completed character, 19
complexity, 65
compliance, 15, 35, 78, 85, 86, 133, 134, 135, 147
Comprehensive Environmental Response, Compensation, and Liability Act (CERCLA), 4, 15, 21, 106
conference, 92, 105
configuration, 56
congress, 1, 5, 32, 48, 49, 68, 85, 102, 108, 129, 137, 138, 148
connectivity, 93
consensus, 89
consent, 12
constituents, 36, 63, 90, 91, 114, 130, 134, 136
construction, 15, 71, 76, 80, 85, 94, 104, 109, 113, 115, 133
consulting, 77
consumption, 6, 11, 36, 79, 112
contaminant, 15, 67, 79, 91, 92, 93, 131, 132
contaminated sites, 12, 13, 32, 40, 50, 64, 65, 98
contaminated soil, 12, 14, 15, 18, 68, 70, 84, 103, 104, 108, 119, 122, 123
contaminated soils, 12, 14, 15, 70, 84, 108
contaminated water, 35, 36, 69, 71, 80, 102, 103, 112
contingency, 29, 55, 58, 59
contour, 93
conversations, 139, 140
cooking, 43
cooperation, 2, 17, 35, 138
cooperative agreements, 13, 67
coordination, 18, 47, 66, 70, 71, 76, 84, 109, 118
correlation, 59
cost, 2, 5, 24, 30, 31, 33, 34, 36, 37, 39, 48, 49, 50, 53, 56, 57, 58, 59, 61, 62, 63, 64, 66, 67, 68, 70, 71, 72, 75, 76, 77, 78, 80, 83, 84, 89, 93, 109, 122, 124, 129, 133, 135, 146, 148
Court of Appeals, 65
covering, 90, 130
crude oil, 64
crust, 21, 63
cultural practices, 22
culture, 6
cure, 66

D

danger, 37
data analysis, 132
data collection, 132
data set, 25
database, 75, 81, 82, 117, 118, 140, 141
deaths, 20
decision makers, 2, 31, 33, 34, 47, 48
decision-making process, 40
decontamination, 64
defects, 142
deficiencies, 66
Department of Defense, 89
Department of Energy, v, 2, 3, 4, 15, 18, 25, 51, 68, 69, 88, 90, 98, 101, 102, 147
Department of Health and Human Services, 4, 49, 50, 51, 63, 138
Department of Justice, 10, 52, 68, 70, 72, 82, 86, 98, 126, 137, 138
Department of Labor, 138
Department of the Interior, 4, 30, 51, 52, 56, 57
deposits, 8, 9, 10
depth, 72, 90, 129
detection, 17, 131
detonation, 16
developmental milestones, 142
direct cost(s), 25, 52, 65
disappointment, 41
diseases, 17
disposition, 86
distribution, 87, 89, 148
District of Columbia, 65
DOI, 81, 86, 89, 92, 122, 136
DOJ, 98
DOL, 138, 147

draft, 2, 28, 30, 31, 32, 34, 49, 50, 51, 66, 70, 132
drainage, 104, 122, 124
drinking water, vii, 3, 4, 10, 14, 15, 16, 19, 22, 23, 24, 27, 28, 30, 35, 36, 37, 38, 39, 43, 45, 63, 67, 70, 73, 79, 80, 82, 90, 91, 93, 101, 103, 104, 106, 112, 113, 114, 118, 130, 131, 132

E

economic development, 30, 45
education, 17, 20, 21, 22, 43, 76, 95, 96, 139, 143
electricity, 7
e-mail, 29, 42
emergency, 12, 18, 38, 117, 118
emergency response, 18, 117, 118
employees, 23, 45, 138
employment, 122, 137
energy, 147
enforcement, 12, 18, 72, 73, 81, 82, 83, 89, 91, 122, 126, 128, 130
engineering, 22, 61, 64, 67, 84, 94
environment, 10, 12, 13, 39, 45, 64, 66, 68, 81, 86, 87, 92, 94, 98, 99, 105, 107, 117, 121, 122, 127, 131, 133, 135, 136, 137
environmental contamination, 108
environmental effects, 6
environmental impact, 70, 94
Environmental Protection Agency, v, 2, 3, 14, 15, 18, 21, 25, 48, 49, 51, 52, 56, 61, 69, 75, 79, 81, 90, 98, 101, 102, 147
epidemiologic, 95
epidemiologic studies, 95
equipment, 9, 18, 76, 109
erosion, 88
evaporation, 9, 94
evidence, 6, 21, 34, 36, 54, 56, 57, 58, 59, 63, 80, 89, 106, 119, 134, 136
examinations, 74, 95, 105, 140
exercise, 87
expenditures, 25, 144
expertise, 94
exposure, viii, 3, 5, 6, 10, 11, 17, 46, 62, 63, 71, 76, 79, 82, 94, 95, 96, 101, 105, 106, 107, 109, 118, 119, 122, 131, 137, 138, 139, 140, 141, 142
extraction, 92, 134
extracts, 10

F

face-to-face interaction, 42
families, vii, 3, 11, 68, 75, 79, 80, 101
family members, 68, 76, 110
fears, 41
federal agency, 5, 18, 19, 20, 40, 47, 52, 53, 136
federal funds, 33, 36, 38, 53, 67
federal government, 6, 10, 12, 17, 33, 40, 45, 64, 65, 67, 89, 138
Federal Register, 143
fencing, 42, 104, 105, 106, 119, 131, 141, 148
financial, 10, 18, 20, 30, 40, 56, 62, 103, 106
financial incentives, 10
financial resources, 106
fiscal year 2009, 23, 24
flexibility, 55
food, 94
food chain, 94
force, 55
formation, 91
foundations, 75, 108
funding, vii, 1, 2, 15, 22, 23, 24, 31, 32, 33, 35, 36, 38, 39, 40, 41, 42, 45, 48, 49, 52, 53, 62, 64, 65, 67, 70, 77, 80, 82, 83, 86, 87, 95, 103, 113, 130, 138, 143, 144, 146
funds, 15, 16, 22, 23, 24, 36, 38, 39, 48, 53, 62, 65, 71, 73, 80, 81, 86, 103, 105, 113, 117, 127, 133, 135, 139, 144, 148

G

gamma radiation, 11, 77, 88, 106, 108, 117, 118, 119
GAO, vii, 1, 2, 3, 7, 9, 12, 13, 16, 18, 19, 20, 21, 25, 54, 55, 56, 57, 58, 59, 61, 63, 64, 66, 67, 68
Geographic Information System, 117
geology, vii, 101, 105
GIS, 75, 81
GPS, 118
grants, 25, 64
gravity, 113
grazing, 42
gross domestic product, 52
groundwater, vii, viii, 4, 5, 9, 10, 13, 15, 16, 19, 25, 27, 28, 31, 33, 35, 37, 51, 63, 64, 65, 66, 67, 69, 70, 72, 73, 79, 85, 86, 87, 88, 90, 91, 92, 93, 94, 98, 101, 102, 105, 128, 130, 131, 132, 133, 134, 135, 136, 137
guidance, 31, 56, 57, 70
guidelines, 113, 146

H

Hawaii, 68
hazardous materials, 11
Hazardous Substance Superfund Trust Fund, 12
hazardous substances, 12, 64

hazardous waste, 14, 22, 91
hazards, 4, 9, 11, 13, 14, 46, 81, 95, 117
HDPE, 124
health, vii, 1, 3, 4, 5, 6, 11, 13, 16, 17, 18, 19, 20, 22, 23, 36, 43, 45, 61, 62, 63, 64, 65, 68, 69, 70, 71, 73, 76, 78, 79, 82, 83, 86, 91, 94, 95, 96, 97, 102, 103, 104, 105, 106, 107, 109, 112, 113, 117, 138, 139, 140, 141, 142, 143, 144, 146
health and environmental effects, 4
Health and Human Services, 49, 68
health care, 16, 20, 62, 95, 139
health care professionals, 139
health condition, 5, 19, 20, 22, 61, 65, 70, 73, 95, 96, 97, 105, 138
health education, 139, 142, 143
health effects, vii, 1, 3, 4, 6, 11, 17, 21, 43, 46, 79, 95, 142
health information, 16
health problems, 69, 71, 79, 82, 102, 104, 113, 117, 139
health researchers, 142
health risks, 18, 68, 70, 91, 102, 104, 105, 112, 113
health services, 16, 138
health status, 95
heavy metals, 9, 139
HHS, 89, 138, 139, 143, 148
history, 109, 141
homes, vii, 3, 4, 6, 10, 11, 15, 16, 18, 23, 24, 35, 36, 42, 43, 47, 61, 62, 63, 67, 71, 75, 76, 78, 80, 81, 101, 103, 104, 106, 108, 109, 113, 115, 119, 122, 143, 148
Hopi reservations, vii, 1, 6, 52
hotel, 76
House, 4, 31, 34, 63, 68, 69, 102, 147
House of Representatives, 147
House Report, 147
housing, 18, 110, 122
Housing and Urban Development, 20, 24, 104
HUD, 104, 106, 113, 143
human, 11, 12, 13, 20, 36, 38, 64, 66, 68, 71, 79, 86, 87, 94, 105, 106, 107, 112, 122, 127, 131, 133, 135
human exposure, 79
human health, 11, 12, 13, 20, 38, 64, 66, 68, 71, 79, 86, 87, 94, 105, 106, 107, 112, 122, 127, 131, 133, 135

I

identification, 54, 107
images, 9
improvements, 98, 137
income, 64, 68
Indian reservation, 12
Indians, 5, 16, 20, 64, 70, 71, 73, 80, 95, 96, 105, 138
individuals, 10, 17, 53, 68, 70, 95, 102, 105, 112, 113, 137, 139, 140, 141
industry, 5, 10, 30, 33, 137, 139
infant mortality, 20
inflation, 52, 58
information sharing, 103, 107
infrastructure, 10, 15, 23, 24, 35, 38, 64, 83, 113
ingestion, 11
initiation, 19
in-kind services, 141
internal controls, 53
internship, 22
irrigation, 73, 91, 92, 93, 131
issues, 4, 15, 17, 27, 34, 40, 41, 42, 43, 45, 46, 53, 55, 73, 77, 78, 86, 95, 96, 98, 103, 117, 118, 119, 136, 140, 142, 146

J

job creation, 10, 45
job skills, 144
job training, 22, 45, 122, 144
job training programs, 45
jurisdiction, 10, 43, 79
justification, 55, 93

K

kidney, viii, 3, 20, 37, 101, 119, 140, 141

L

laboratory tests, 139
landscape, 76
languages, 42
laws, 10, 53, 68
leaching, 9, 94
lead, 14, 16, 27, 33, 48, 70, 72, 82, 83, 89, 126
legislation, 33, 75, 83
leukemia, 11
light, 28, 49, 67
livestock, 5, 6, 11, 43, 71, 76, 78, 79, 112, 113
loans, 67
local community, 139
local government, 6, 7, 36, 53
lung cancer, viii, 3, 11, 101, 119

M

magnitude, 34, 65
majority, 137, 148
management, 2, 28, 29, 30, 31, 42, 48, 50, 54, 55, 56, 57, 64, 66, 87, 94, 122, 136, 139, 140
mapping, 55, 117
materials, 4, 9, 10, 11, 13, 16, 21, 33, 41, 42, 44, 53, 54, 57, 63, 72, 75, 76, 77, 86, 87, 89, 91, 104, 106, 108, 109, 119, 122, 129, 138
matrix, 59
measurement(s), 55, 117, 119
media, 42
medical, 20, 95, 105, 113, 139, 140, 141
metals, 90, 130, 138
methodology, 6, 57, 94
Mexico, 4, 6, 9, 10, 20, 41, 52, 53, 63, 65, 67, 72, 79, 80, 84, 86, 96, 99, 104, 106, 107, 113, 118, 121, 122, 134, 135, 137, 139, 140, 141, 142, 143, 144, 145, 147, 148
migration, 15, 91, 92, 93, 104, 122, 124, 131
minimum price, 10
mission, 62, 141
modifications, 29
moratorium, 10

N

National Defense Authorization Act, 34, 40
National Environmental Policy Act (NEPA), 85, 133
National Institutes of Health, 68, 105
National Priorities List, 3, 13, 18, 20, 84, 117, 118
Navajo people, vii, 1, 3, 6, 10, 11, 17, 20, 36, 40, 42, 44, 45, 47, 95, 101
Navajo reservation, vii, 1, 2, 3, 4, 5, 6, 7, 8, 9, 10, 11, 13, 14, 15, 16, 17, 20, 24, 31, 34, 35, 36, 38, 40, 41, 42, 45, 47, 48, 49, 50, 52, 53, 61, 62, 63, 64, 65, 66, 67, 139
negotiating, 146
networking, 144
nitrates, 130
nonprofit organizations, 6
NPL, 3, 13, 39, 41, 64, 81, 82, 84
NRC, 3, 16, 19, 20, 25, 27, 34, 41, 49, 52, 63, 64, 65, 67, 70, 73, 84, 86, 87, 89, 96, 97, 98, 99, 102, 104, 105, 107, 121, 122, 123, 130, 133, 134, 135, 136, 137, 144, 146
Nuclear Regulatory Commission(NRC), v, 3, 4, 16, 18, 25, 49, 52, 69, 73, 84, 98, 101, 102
nuclear weapons, vii, 1, 3, 9, 17, 45, 137, 138

O

occupational health, 141
Office of Management and Budget, 3, 20, 49, 52, 56
officials, vii, 1, 2, 5, 14, 17, 21, 22, 23, 24, 25, 26, 27, 28, 29, 30, 32, 33, 34, 35, 36, 37, 38, 39, 40, 41, 42, 43, 44, 45, 46, 48, 52, 53, 56, 57, 61, 62, 63, 64, 65, 66, 67, 75, 76, 77, 104, 107, 112, 118
oil, 91
OMB, 3, 20, 65
Omnibus Appropriations Act,, 148
On-Scene Coordinator, 89
operations, vii, 3, 10, 63, 65, 76, 98, 101, 137
opportunities, vii, 1, 2, 5, 6, 22, 38, 44, 45, 46, 52, 53, 55, 86, 107, 143, 144
organ, 144
organize, 46, 48, 64
outreach, 2, 15, 21, 22, 26, 30, 40, 41, 42, 43, 44, 45, 46, 47, 49, 50, 62, 71, 73, 76, 80, 99, 104, 106, 107, 113, 142, 143, 146
oversight, 16, 39, 105, 122, 129, 132, 133, 136, 146
ownership, 7, 8
ox, 63

P

participants, 20, 44, 99, 140, 141, 143, 144
pathways, 10, 11, 15, 64, 91, 94, 131
penalties, 29, 48, 50, 64, 66
photographs, 117, 118
physical structure, 43
physics, 139
pipeline, 92
plants, 11
playing, 11
policy, 72, 82, 107, 112, 126, 144
policy issues, 107, 144
policymakers, 47
pollutants, 68
ponds, 9, 94
poor performance, 29
population, 6, 39, 78, 83, 112, 139
population density, 83
potassium, 21, 63
poverty, 6
power plants, 9, 10
preparation, 72, 89, 129
president, 45
prevention, 17
price index, 52

processing sites, vii, 1, 5, 11, 13, 14, 15, 16, 19, 22, 24, 25, 33, 35, 37, 40, 41, 42, 46, 50, 62, 67, 72, 85, 86, 133, 135, 138
professionals, 141
program staff, 140
project, 2, 10, 19, 20, 23, 24, 26, 27, 28, 29, 30, 31, 39, 48, 49, 50, 51, 53, 54, 55, 56, 57, 58, 66, 68, 81, 84, 93, 104, 105, 113, 130, 141, 144, 146, 148
protection, 87, 105, 122, 131
public affairs, 42
public education, 80
public health, vii, 45, 69, 70, 79, 92, 98, 99, 103, 112, 136, 137
pumps, 113
PVC, 92

R

radiation, vii, 3, 4, 11, 12, 15,17, 18, 19, 20, 21, 36, 59, 60, 61, 63, 65, 69, 67, 68, 71, 75, 76, 77, 85, 88, 89, 90, 96, 98, 103, 104, 106, 108, 109, 112, 117, 119, 124, 137, 138, 139, 146, 147, 148
radio, 104, 113
radioactive waste, 9, 10, 77, 91, 129
radium, vii, 9, 10, 21, 63, 84, 90, 101, 121, 130
radius, 113, 148
radon, 9, 10, 11, 67, 75, 76, 77, 108
realism, 55
receptors, 91, 93, 94, 131
recommendations, 2, 33, 49, 50, 131
recovery, 63, 70, 73, 84, 98, 136, 137, 146
recruiting, 20, 142
reform, 4, 31, 34, 63, 69, 102, 147
Registry, 3, 5, 16, 25, 52, 62, 96, 105
regulations, 41, 68, 78
regulatory framework, 98, 137
regulatory oversight, 135
reimburse, 76, 122
reliability, 52, 57
remedial actions, 4, 12, 13, 15, 30, 33, 35, 38, 40, 48, 64, 66, 70, 82, 89, 94, 99, 110, 132, 136
remediation, 13, 16, 26, 35, 37, 51, 65, 66, 72, 73, 75, 76, 77, 83, 85, 86, 87, 88, 89, 90, 91, 94, 98, 99, 103, 104, 105, 109, 122, 128, 129, 130, 132, 133, 134, 135, 136, 143
removal actions, 12, 17, 18, 24, 33, 36, 37, 38, 39, 61, 65, 67, 76, 81, 82, 121, 122
reprocessing, 121
requirements, 10, 29, 31, 34, 39, 48, 56, 65, 66, 98, 112, 137, 146
researchers, 6, 11, 53, 62, 141, 142
Reservations, 7
Residential, 110
resolution, 10
resource allocation, 2, 47, 49
resources, 2, 15, 22, 23, 24, 28, 31, 38, 39, 46, 48, 49, 50, 52, 54, 55, 68, 82, 86, 93, 99, 103, 107, 108, 119, 126, 136, 138, 139, 146
response, 2, 5, 17, 18, 28, 31, 39, 51, 64, 65, 71, 72, 73, 76, 80, 82, 83, 89, 91, 102, 104, 107, 109, 119, 124, 128, 130, 131, 146
restitution, 137
restoration, 13, 63, 64, 94
revenue, 64
risk(s), 11, 13, 17, 18, 32, 34, 36, 37, 38, 42, 51, 54, 55, 58, 59, 61, 63, 64, 66, 70, 71, 75, 76, 78, 79, 82, 84, 92, 94, 98, 102, 103, 106, 108, 109, 112, 117, 127, 137, 146
risk assessment, 94
routes, 94, 139
rules, 57
rural areas, 79

S

SACE, 79
safety, 11, 27, 70, 98, 99, 136, 137, 146
scarcity, 30
science, 22
scope, vii, 1, 2, 4, 5, 17, 28, 30, 31, 32, 33, 34, 35, 36, 41, 43, 44, 47, 48, 49, 51, 52, 53, 57, 61, 73, 83, 87, 90, 91, 92, 102, 103, 104, 106, 112, 119, 146
second language, 42
sediments, 75
seeding, 94
selenium, viii, 101, 130, 134
Senate, 68
sensitivity, 44, 58
sequencing, 54, 55
services, 17, 80, 81, 95, 138, 139, 141
settlements, 18, 109
sheep, 6, 12
showing, 117, 119
signs, 9, 36, 42, 104, 112, 119
site cleanup, 12, 84, 121, 130, 144
small communities, 39
sodium, 63
software, 55, 83
solid waste, 90, 130
solution, 30
solvents, 138
species, 72, 90, 129
spectroscopy, 131
spending, 24, 36, 47, 62, 63
Spring, 76, 77

stabilization, 89, 104, 106
staff members, 42, 62, 99
stakeholders, vii, 1, 2, 6, 22, 30, 31, 33, 34, 41, 43, 45, 47, 53, 62, 94, 107, 132, 136, 143
state(s), vii, 6, 7, 8, 42, 51, 57, 64, 67, 86, 101, 118, 133, 135
statutes, 6, 11
steel, 63
stock, 113, 148
storage, 17, 109
structure, 17, 21, 54, 57, 59, 65, 71, 75, 76, 77, 78, 106, 109, 119
sulfate, 134
Superfund, 11, 12, 13, 14, 18, 22, 23, 36, 38, 64, 67, 71, 72, 73, 75, 76, 77, 79, 80, 82, 83, 84, 89, 91, 109, 112, 121, 126, 128, 130, 144
surveillance, 5, 15, 19, 20, 26, 27, 41, 64, 72, 73, 86, 96, 105, 122, 130, 133, 135, 138
survivors, 138

T

target, 23, 26, 27, 30, 80, 104, 109
Task Force, 107
taxes, 64
technical assistance, 80, 136, 143
technical comments, 49
techniques, 55, 71, 81
technology, 22
telephone(s), 7, 42, 53
temporary housing, 76, 122
Tennessee Valley Authority, 85
territorial, 10
testing, 17, 61, 67, 92, 94, 103, 131, 132, 137
theft, 43
thorium, 21, 63
threats, 12, 19, 26, 84, 94
time frame, vii, 1, 2, 5, 26, 31, 32, 33, 34, 36, 47, 48, 49, 52, 53, 61, 67, 107, 132
Title I, 64, 85, 86, 87, 98, 136
Title II, 64, 87
total costs, 51
toxic substances, 138
toxic waste, 38
toys, 36
tracks, 36
training, 20, 22, 45, 105, 139, 143, 144
transmission, 104, 113, 115
transport, 10, 37, 148
transportation, 10, 67, 77, 94, 104
Treasury, 64
treatment, 19, 22, 31, 40, 62, 67, 68, 70, 72, 87, 88, 89, 94, 105, 128, 133, 135, 136
tribal lands, 4
tribal leaders, vii, 1, 28
tribal officials, 4, 5, 52, 53, 118
Trust Fund, 12, 13, 38, 64, 67

U

U.S. Army Corps of Engineers, 79, 83
U.S. Department of Labor, 147
U.S. Geological Survey, 15, 90
unique features, 56
United States, v, 1, 6, 8, 9, 20, 33, 34, 63, 64, 65, 68, 102, 108, 122, 134, 136
updating, 54
uranium contamination, vii, 1, 2, 3, 4, 5, 6, 11, 14, 15, 16, 21, 22, 31, 32, 34, 35, 36, 37, 38, 40, 41, 44, 45, 46, 47, 48, 49, 50, 52, 53, 67, 70, 91, 102, 103, 106, 107, 132, 136, 144, 146
uranium milling sites, vii, 69, 70, 73, 74, 85, 86, 105, 108, 133, 135
uranium mining, vii, 3, 4, 6, 8, 9, 10, 11, 14, 36, 40, 45, 47, 51, 53, 64, 67, 68, 69, 70, 75, 79, 81, 119, 134, 137, 139
urban, 7
USA, 64, 127
USGS, 72, 89, 92, 93, 128, 132, 136

V

validation, 132
vanadium, 63, 90, 130
vandalism, 43
vapor, 91, 132
vegetation, 132
videos, 46

W

Washington, 8, 63, 64, 66, 67, 68, 147
waste, 5, 9, 11, 15, 17, 19, 27, 32, 33, 35, 36, 37, 61, 66, 67, 71, 75, 79, 81, 83, 90, 91, 94, 104, 107, 108, 112, 116, 117, 121, 122, 124, 130, 131, 132, 143, 144, 148
waste disposal, 33, 94, 122
water quality, 35, 67, 92, 131
water quality standards, 35, 67
water sources, vii, 1, 3, 4, 7, 16, 19, 22, 23, 30, 35, 36, 43, 69, 70, 71, 78, 79, 80, 82, 101, 102, 103, 109, 112, 113, 114, 118
water supplies, 19, 27, 69, 71, 72, 80, 81, 91, 92, 93, 102, 103, 106, 107, 112, 130, 146
weapons, vii, 3, 101, 137, 138

web, 93
wells, 5, 7, 10, 19, 23, 27, 28, 30, 43, 65, 71, 73, 78, 79, 90, 91, 92, 93, 104, 105, 112, 113, 130, 131, 132, 137, 148
wood, 4
wool, 6
workers, 20, 22, 76, 86, 137, 138
World War I, vii, 101